1975

Introduction to Optimization Methods

CHAPMAN AND HALL
MATHEMATICS SERIES

Edited by Professor R. Brown.
Head of the Department of Pure Mathematics,
University College of North Wales, Bangor,
and Dr M.A.H. Dempster.
University Lecturer in Industrial Mathematics
and Fellow of Balliol College, Oxford

A Preliminary Course in Analysis
R.M.F. Moss *and* G.T. Roberts

Elementary Differential Equations
R.L.E. Schwarzenberger

A First Course on Complex Functions
G.J.O. Jameson

Rings, Modules and Linear Algebra
B. Hartley *and* T.O. Hawkes

Regular Algebra and Finite Machines
J.H. Conway

Complex Numbers
W.H. Cockcroft

Galois Theory
Ian Stewart

Topology and Normed Spaces
G.J.O. Jameson

Introduction to Optimization Methods

P.R. ADBY

Lecturer in Electrical Engineering
University of London, King's College

and

M.A.H. DEMPSTER

University Lecturer in Industrial Mathematics and Fellow of Balliol College, Oxford

CHAPMAN AND HALL
London

A HALSTED PRESS BOOK
JOHN WILEY & SONS
New York

First Published 1974
by Chapman and Hall Ltd
11 New Fetter Lane, London EC4P 4EE

Set by E.W.C. Wilkins Ltd, London and Northampton
Printed in Great Britain by
Whitstable Litho, Straker Brothers Ltd

Library of Congress Cataloging in Publication Data

Adby, Paul Raymond
Introduction to optimization methods.

(Chapman and Hall mathematics series)
Bibliography: p.
1. Mathematical optimization. I. Dempster, Michael Alan Howarth, joint author. II. Title.
QA402.5.A4 515 74–4109
ISBN 0–470–08830–X

Preface

During the last decade the techniques of non-linear optimization have emerged as an important subject for study and research. The increasingly widespread application of optimization has been stimulated by the availability of digital computers, and the necessity of using them in the investigation of large systems.

This book is an introduction to non-linear methods of optimization and is suitable for undergraduate and post-graduate courses in mathematics, the physical and social sciences, and engineering.

The first half of the book covers the basic optimization techniques including linear search methods, steepest descent, least squares, and the Newton-Raphson method. These are described in detail, with worked numerical examples, since they form the basis from which advanced methods are derived.

Since 1965 advanced methods of unconstrained and constrained optimization have been developed to utilise the computational power of the digital computer. The second half of the book describes fully important algorithms in current use such as variable metric methods for unconstrained problems and penalty function methods for constrained problems. Recent work, much of which has not yet been widely applied, is reviewed and compared with currently popular techniques under a few generic main headings.

Chapter 1 describes the optimization problem in mathematical form and defines the terminology used in the remainder of the book. Chapter 2 is concerned with single variable optimization. The main algorithms of both search and approximation methods are developed in detail since they are an essential part of many multi-variable methods. Chapter 3 is devoted to the basic multi-variable methods of optimization and includes the fundamental first and second order gradient methods and the simple direct search methods. Chapter 4 considers advanced gradient methods and advanced search methods for unconstrained optimization. Minimax polynomial approximation is also included as an introduction to curve fitting and in contrast to least squares. Chapter 5 is concerned with constrained optimization, beginning with a description of the Kuhn-Tucker conditions for a constrained optimum.

For students of mathematics, numerical optimization techniques form an important complement to theoretical considerations of applied analysis in the fields of convexity, mathematical programming, calculus of variations, optimal control and function approximation. Optimization methods are of course basic to operations research and management science. They are also relevant to least squares and maximum likelihood methods for statistical problems arising in econometrics and in the physical and life sciences. For engineering students, it has become necessary to introduce optimization at the undergraduate level in computer aided design courses.

In an attempt to make this work intelligible to such a diverse readership, we have made an effort to use a rigorous mathematical notation which is as simple and interdisciplinary as possible. We have also refrained from being over-pedantic in distinguishing between numerical and mathematical, or local and global, optima. Precise meanings should be clear from the context. References to required general mathematics are to widely available texts, rather than to the most modern or elegant. Although a self-contained development of mathematical results required for an understanding of the methods is given, in a work of this size and scope one must be content with references to the research literature for the proofs of convergence results. A feature of this book is the large collection of references from which the material was compiled. No flow charts are included because we hope that the subject will serve, at least for some readers, as a motivation for learning computer programming.

This book has had the benefit of use in manuscript form by students of engineering at King's College and of mathematics, economics and management studies at Oxford; we are grateful for their helpful comments. We would like to thank R. Fletcher, L. Fox and M.J.D. Powell for reading drafts of Chapters 4 and 5. Their detailed comments materially improved the manuscript. Finally, we would like to express our gratitude to Miss D. Orsanic for drawing the figures, to Mrs Mary Bugge and Mrs Valerie Saunders for typing the text, to Miss M.S. David for checking the proofs and to M.A.H.D.'s wife, Ann Lazier, for help in constructing the index.

P.R. ADBY
M.A.H. DEMPSTER

London
Oxford
December 1973

Contents

CHAPTER ONE

The optimization problem

1.1. Introduction

The requirement for methods of optimization arises from the mathematical complexity necessary to describe the theory of systems, processes, equipment, and devices which occur in practice. Even quite simple systems must sometimes be represented by theory which may contain approximations, by parameters which change with time, or by parameters that vary in a random manner. For many reasons the theory is imperfect, yet it must be used to predict the optimum operating conditions of a system such that some performance criterion is satisfied. At best such theory can predict only that the system is near to the desired optimum. Optimization methods are then used to explore the local region of operation and predict the way that the system parameters should be adjusted to bring the system to optimum.

In an industrial process, for example, the criterion for optimum operation is often in the form of minimum cost, where the product cost can depend on a large number of interrelated controlled parameters in the manufacturing process. In mathematics the performance criterion could be, for example, to minimize the integral of the squared difference between a specified function and an approximation to it generated as a function of the controlled parameters. Both of these examples have in common the requirement that a single quantity is to be minimized by variation of a number of controlled parameters. In addition there may also be parameters which are not controlled but which can be measured, and possibly some which cannot even be measured.

Before the advent of high speed computers this type of problem was insoluble for systems in which the number of parameters was large. Human judgement alone is often unable to optimize systems with only three variables. The use of computers has led to the widespread development and use of the theory of optimization and a number of algorithms are now available so that systems of more than one hundred variables can be treated. In special cases, particularly for linear performance criteria, much larger problems can be solved.

The importance of optimization lies not in trying to find out all about a system but in finding out, with the least possible effort, the best way to adjust the system. If this is carried out well, systems can have a more economic and improved design, they can operate more accurately or at less cost, and the system designer will have a better understanding of the effects of parameter interaction and variation on his design.

This book is an introduction to methods for computing solutions to the non-linear optimization problem, with emphasis on basic methods which are used in modified form as part of more complex methods currently available.

1.2 Problem definition

Minimization

The basic mathematical optimization problem is to minimize a scalar quantity E which is the value of a function of n system parameters $x_1, x_2, \ldots, x_n$. These variables must be adjusted to obtain the minimum required, i.e. to

$$\text{minimize } E = f(x_1, x_2, \ldots, x_n). \tag{1.1}$$

The optimization problem is formulated here as a minimization problem. This is not restrictive since almost all problems can be put in this form. In particular a maximum of a function can be determined by a minimization method since

$$\text{maximum } f(x) = -\text{minimum } \{-f(x)\}. \tag{1.2}$$

The value E of f embodies the design criteria of the system into a single number which is often a measure of the difference between the required performance and the actual performance obtained. The function f is referred to as the *objective function* whose value is the quantity which is to be minimized.

The n system parameters will be manipulated as the column vector $\mathbf{x}$. The transpose of $\mathbf{x}$ is given in matrix form by

$$\mathbf{x}^{\mathrm{T}} = [x_1 x_2 \ldots x_n] \tag{1.3}$$

where ^{T} signifies the transpose of a matrix. In this work the co-ordinates of $\mathbf{x}$ will take successive values as adjustments are made to their values during the optimization process. Each set of adjustments to these variables is termed an *iteration* and in general a number of iterations are required before an optimum is satisfactorily approximated. After i iterations the value of E will be given by E_i and the value of $\mathbf{x}$ will be $\mathbf{x}_i$. In the iterative processes or *algorithms* to be studied, a first estimate of the parameter values must be supplied as a starting point for the search for minimum. Since at that stage no iterations have taken place, the first estimate of $\mathbf{x}$ will be denoted by $\mathbf{x}_0$ and the resulting objective function value will be E_0. The final optimal values obtained will usually be denoted by $\mathbf{x}_{\mathrm{min}}$ and E_{min} respectively. Changes in the parameter values will be denoted by the vector $\boldsymbol{\Delta}\mathbf{x}$. Its transpose is given by

$$\boldsymbol{\Delta}\mathbf{x}^{\mathrm{T}} = [\Delta x_1 \Delta x_2 \ldots \Delta x_n]. \tag{1.4}$$

Local and global minima

The determination of the parameters $\mathbf{x}_{\mathrm{min}}$ which give a minimum value E_{min} of the objective function f is the object in optimization. E_{min} is the lowest possible value of E for all combinations of values of the variables $\mathbf{x}$. A point $\mathbf{x}_{min}$ which gives this lowest possible value of f is termed a *global minimum*. In general it need not be unique.

In practice it is very difficult to determine if the minimum obtained by a numerical process is a global minimum or not. In most circumstances it can only be said that the minimum obtained is a minimum within a local area of search. (For simplicity we are here ignoring the problem of limits to numerical accuracy.) The point $\mathbf{x}_{min}$ is therefore termed a *local minimum*. Again it need not be unique even locally, but since a given convergent numerical process implemented on a particular computer should always converge to the same point from a given initial value $\mathbf{x}_0$, the local minimum will be referred to throughout this book as *the* minimum without confusion or essential loss of generality. A particular function may, of course, possess several local minima. One of these will be the global minimum but it is usually impossible to determine if a local minimum is also the global minimum unless all minima are found and evaluated.

Gradients

Some optimization methods require gradient information about the objective function given by $E = f(x)$. This is obtained in the form of first and second order partial derivatives of f with respect to the n parameters. The *Jacobian gradient vector* $\mathbf{g}$ is defined as the transpose of the gradient vector ∇f which is a row matrix of first order partial derivatives. The transpose of $\mathbf{g}$ is therefore given by

$$\mathbf{g}^{\mathrm{T}} = \nabla f = \left[\frac{\partial f}{\partial x_1}\ \frac{\partial f}{\partial x_2} \cdots \frac{\partial f}{\partial x_n}\right]. \tag{1.5}$$

The $n \times n$ symmetric matrix of second order partial derivatives of f is known as the *Hessian matrix* and is denoted by $\mathbf{H}$ where

$$\mathbf{H} = \begin{bmatrix} \dfrac{\partial^2 f}{\partial x_1^2} & \dfrac{\partial^2 f}{\partial x_1 \partial x_2} & \cdots & \dfrac{\partial^2 f}{\partial x_1 \partial x_n} \\ \dfrac{\partial^2 f}{\partial x_2 \partial x_1} & & & \vdots \\ \vdots & & & \vdots \\ \dfrac{\partial^2 f}{\partial x_n \partial x_1} & \cdots & \cdots & \dfrac{\partial^2 f}{\partial x_n^2} \end{bmatrix} \tag{1.6}$$

The availability of the Jacobian vector and Hessian matrix simplifies optimization procedures and can sometimes allow solution of problems with a larger number of variables. However, it is usually necessary for more general methods of optimization to proceed without gradient information since derivatives may be uneconomic or impossible to compute, and in some cases may not even exist.

Constraints

In many practical optimization problems there are constraints on the values of some of the parameters which restrict the region of search for the minimum. A common constraint on the variable x_i is in the form of the inequality $x_{\mathrm{L},i} \leqslant x_i \leqslant x_{\mathrm{U},i}$ where $x_{\mathrm{L},i}$ and $x_{\mathrm{U},i}$ are fixed lower and upper limits to x_i. More generally 'inequality constraints' are formulated to specify functional relationships of the parameters involved in the constraint. Most inequality constraints can be fitted into the form

$$g(x_1, x_2, \ldots, x_n) \leqslant 0. \tag{1.7}$$

For the simple case of upper and lower limits, the expression $x_{\mathrm{L},i} \leqslant x_i \leqslant x_{\mathrm{U},i}$ is replaced by the two expressions

$$x_i - x_{\mathrm{U},i} \leqslant 0 \tag{1.8}$$

$$x_{\mathrm{L},i} - x_i \leqslant 0. \tag{1.9}$$

The region of search in which the constraints are satisfied is termed the *feasible region*, while the region in which constraints are not satisfied is termed the *non-feasible* or *infeasible region.*

It is also possible to optimize systems in which the parameters are constrained to specific functional relationships. This type of constraint can be formulated in a similar way to inequality constraints and then equations of the form

$$h(x_1, x_2, \ldots, x_n) = 0 \qquad (1.10)$$

are obtained.

Optimization with constraints is very much more difficult than unconstrained optimization and a great deal of effort has been expended to reformulate constrained problems so that constraints are avoided. For this reason constrained optimization is not considered until the final chapter.

Convergence

Two problems of convergence of the iterative process of a given optimization method arise and it is necessary to answer the following questions before different techniques can be compared.

(i) *Has the value E of the objective function converged to a minimum and is this minimum a global minimum*?

(ii) *What was the speed of convergence*?

The value of E for successive iterations will in most cases be the only available guide to the progress of the minimization When E does not reduce over a number of iterations progress has clearly stopped and some kind of minimum has been reached. The geometric interpretation of this will be deferred until Section 1.4, but to summarize: *it is almost impossible to predict if the minimum reached is the global minimum.* None of the iterative techniques described will guarantee convergence to the global minimum when several minima exist. It seems unlikely that any non-linear optimization technique can do so in general. Even extensive testing of all minima found is not a complete answer, since a given technique may never converge to the global minimum of certain functions.

The relative speed of convergence is usually assessed by the number of evaluations of the function f necessary to reduce E by a specified amount, since the time required for a given computer to reach a solution is largely made up of function evaluations.

Sampled data

In most scientific and engineering problems the quantity to be minimized is a function not only of controlled parameters $x_1, x_2, \ldots, x_n$ but also of one or more independent variables (for example, time or position). The objective function f then takes a series of values as the independent variables vary. Physical measurement of such an objective function can only be entered into a digital computer as a series of values at specific sample values of the independent variables. In addition the number of samples may be limited, due either to practical difficulties in making measurements, or to disturbance of the system by the measurement device.

A similar situation exists in mathematical problems when independent variables are involved (for example when optimizing the choice of a function of a prescribed type to approximate a given function) since functions can only be evaluated by the computer for specific sets of values of the function arguments. In practice, therefore, such curve fitting and similar processes are handled by the use of a number of sample points sufficient to describe the graphs of the functions involved.

A large number of optimization problems are therefore concerned with functions in which data is only available at a number of sample points. The number of sample points is dependent on the function and may vary from less than ten to several hundreds.

The basic optimization problem in these cases is in a slightly different form to that stated in expression 1.1. The value $\mathbf{E}$ of the objective function is now a vector whose elements are errors at the individual sample points. All elements of $\mathbf{E}$ must then be minimized in a suitable sense simultaneously. If the problem is restated so that the individual errors of E are combined into one scalar quantity, standard methods of optimization can be applied. The new objective is to minimize E where E is a function dependent on the parameters $x_1, x_2, \ldots, x_n$ through functions of the parameters and the independent variables, e.g. in the form

$$E = g\{f(\mathbf{x},\mathbf{t}_1), f(\mathbf{x},\mathbf{t}_2), \ldots, f(\mathbf{x},\mathbf{t}_m)\} \qquad (1.11)$$

where $\mathbf{t}_1, \mathbf{t}_2, \ldots, \mathbf{t}_m$ are vectors of values of the independent variables at the m sample points. The form of the function g is termed the *error criterion* and is discussed below.

Error criteria and weights

The values of the independent variables at each of the m sample points may be incorporated into the function f of $\mathbf{x}$ in equations 1.11 to yield

$$E = g\{f_1(\mathbf{x}), f_2(\mathbf{x}), \ldots, f_m(\mathbf{x})\}. \qquad (1.12)$$

(i) *Least squares*

The function g most often used in curve fitting and other applications (for example in statistics and econometrics) is known as the *least squares* error criterion. This is given by the expression

$$\text{minimize } E = \sum_{i=1}^{m} \{w_i f_i(\mathbf{x})\}^2 \qquad (1.13)$$

where $w_1, w_2, \ldots, w_m$ are termed *weights* or *penalties* and have the effect of emphasizing errors of importance in the formulation of the problem. Weights are also used in cases of several independent variables to equalize the effects of error when the corresponding function values, $f_i(\mathbf{x})$, give errors on a vastly different scale as different independent variables vary.

(ii) *Minimax*

The function g which minimizes the maximum element is known as the *minimax* error criterion, i.e.

$$\text{minimize } E = \max_i |w_i f_i(\mathbf{x})|. \tag{1.14}$$

The main problem with this error criterion is that the error E jumps discontinuously as one value of i changes to another during the course of optimization. Derivatives of E with respect to the parameters are therefore undefined.

(iii) *Other error criteria*

Other error criteria are not considered in this book since they are mainly used in more specialised applications of approximation theory.

Solution of simultaneous equations

A system of non-linear simultaneous equations $f_i(\mathbf{x}) = 0$ can be regarded as defining the functions f_i, $i = 1,2, \ldots, m$ above. Their exact solutions correspond to $E = 0$ in equation 1.12. In general of course an exact solution is not known and the least squares error criterion given in equation 1.13 can be used to improve an initial estimate by minimizing the value of E.

Minimization of the scalar value E is not however recommended, either for the solution of simultaneous equations or in the optimization of sampled systems with independent variables. Instead, least squares methods should be used which minimize the vector valued $\mathbf{E}$ and involve use of the derivatives of each of the elements of $\mathbf{E}$ with respect to each element of $\mathbf{x}$. (Powell's least squares method, given in Section 4.3 is particularly recommended).

1.3. Optimization in one dimension

When the number of system parameters, n, is equal to one, expression 1.1 simplifies to

$$\text{minimize } E = f(x). \tag{1.15}$$

The optimization is therefore a search for a minimum in one dimension only. Since a number of procedures for multi-dimensional optimization reduce to a series of optimizations in one dimension, this problem must be solved efficiently and accurately.

Unimodality and convexity

Typical graphs of the function given by $E = f(x)$ are plotted in Figure 1. The most important common characteristic of the three curves is that within a defined interval of x there is only one minimum value of E. Such 'unimodal' functions have the property that when x is varied incrementally from the minimum x_{min} to the lower limit x_{L} or to the upper limit x_{U} the value of E does not reduce for successive increments in x. More formally, a function of a single argument is *unimodal* when x_{min} is the only value of x for which $f(x) \leqslant f(y)$ for all y in every interval containing x. This description of the minimum of a function avoids reference to derivatives and therefore applies both to continuous and discontinuous functions.

Figure 1a shows a continuous function with continuous derivatives. It demonstrates additionally what is meant by a 'convex' function. In geometrical terms a function is 'strictly convex' if the straight line connecting any two points on its graph lies above the graph. More formally, a function of a single argument is *convex* if for all x and y and all numbers λ between 0 and 1,

$$f(\lambda x + (1-\lambda)y) \leqslant \lambda f(x) + (1-\lambda) f(y). \quad (1.16)$$

If the inequality is strict, f is termed *strictly convex*. The function f is *concave* if $-f$ is convex.

1.4. Optimization in n dimensions

Provided a function is sufficiently differentiable it can be expanded as a Taylor series, see for example Apostal (1957), Chapter 6. If derivatives of all orders do not exist, a series with remainder term may be necessary. The first few terms of the multidimensional Taylor expansion in matrix form are given by

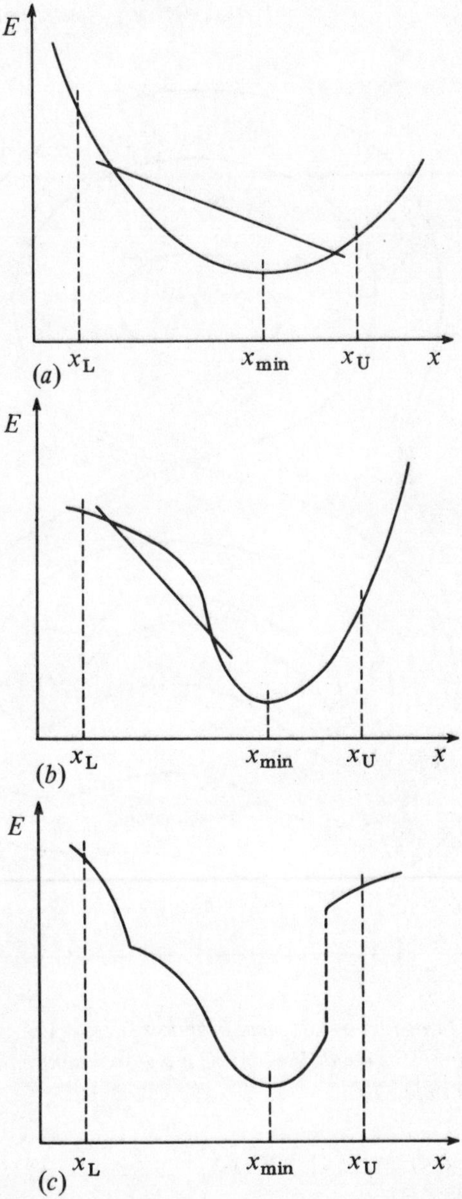

Fig. 1. *Typical unimodal functions. Graph* (*a*). *Continuous convex function. Graph* (*b*). *Continuous non-convex function. Graph* (*c*). *Discontinuous function.*

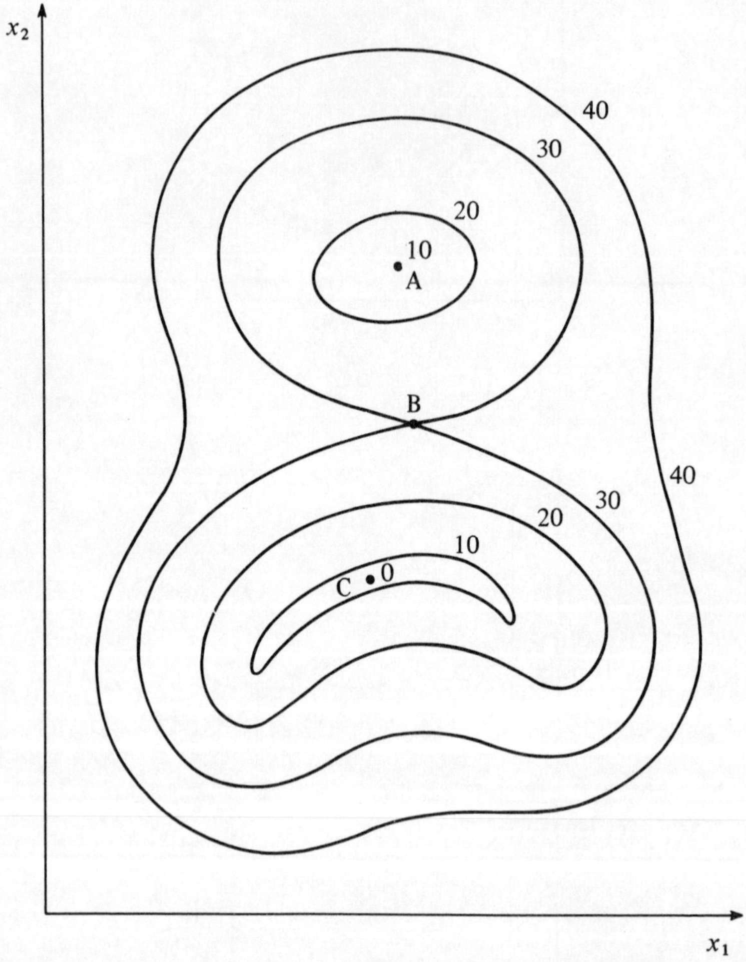

Fig. 2. *Contours of a typical function* $f(x_1, x_2)$. A *is a local minimum.* B *is a saddle point.* C *is a global minimum in a narrow valley.*

$$f(\mathbf{x} + \mathbf{\Delta x}) = f(\mathbf{x}) + \mathbf{g}^T \mathbf{\Delta x} + \tfrac{1}{2} \mathbf{\Delta x}^T \mathbf{H} \mathbf{\Delta x} + \ldots \quad (1.17)$$

where **x**, **Δx**, **g** and **H**, are given by equations 1.3 to 1.6 respectively. (Recall that the superscript T indicates the transpose of a matrix.) If the first derivative exists, then at a

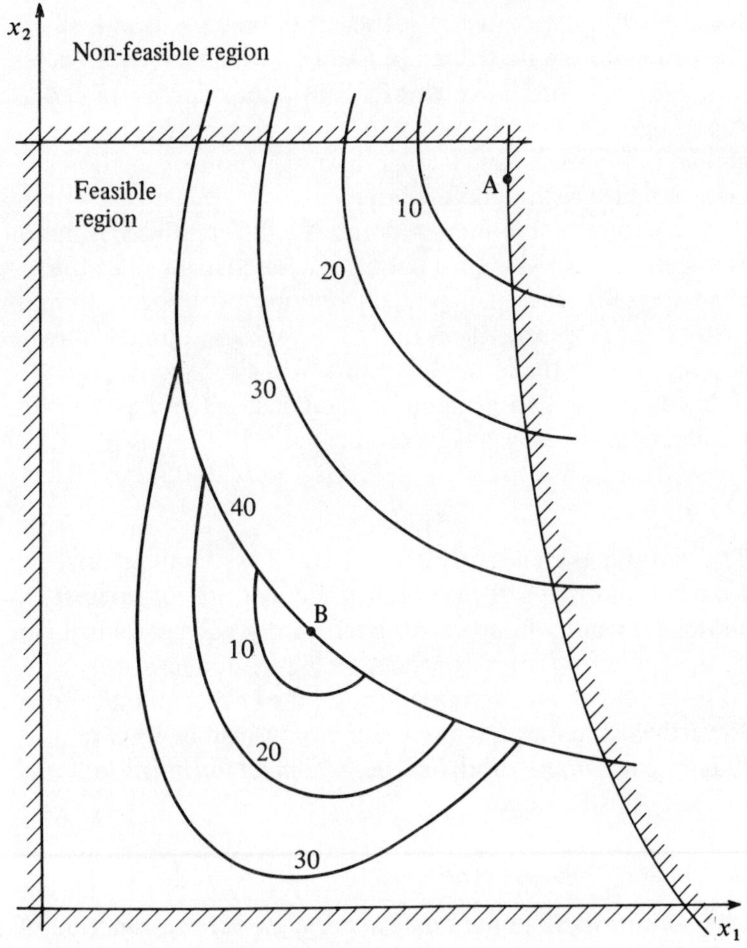

Fig. 3. *Contours of a constrained function $f(x_1, x_2)$ with undefined derivatives.* ⁄⁄⁄⁄ *is the constraint boundary.* A *is the constrained minimum.* B *is a minimum on the path of undefined derivatives.*

local minimum the Jacobian gradient vector **g** has all elements zero. If the second derivatives exist, the Hessian **H** is positive definite at the minimum, see for example Gillespie (1954), Chapter 4, or Apostal (1957), Chapter 7.

Unfortunately, verification of necessary conditions for local minima is very rarely sufficient in practice. Some of the problems are illustrated in Figure 2 which shows a two dimensional contour sketch of a continuous unconstrained function of two variables. Some less obvious difficulties are shown in Figure 3 which concerns a constrained function of two variables with undefined derivatives.

The nature of a minimum point of a differentiable function is one of insensitivity to small parameter changes since the elements of its Jacobian gradient vector must be zero. In most systems this insensitivity is in itself a very good practical reason for seeking the minimum point. When the gradient is not defined, as in a discontinuous system, insensitivity at the minimum point may not be attainable.

Contours

The contour sketches of Figures 2 and 3 are familiar devices. Like contour maps of the countryside, features of interest should be readily interpreted. Each contour is a graphical plot of the function $f(\mathbf{x})$ = constant. In higher dimensions contours are an aid to thought but not of much practical use. Nevertheless, contours in two dimensions can be used to illustrate a number of difficulties which all optimization methods must overcome.

Unimodality and convexity

The definition of a unimodal function given for a function of a single variable in Section 1.3 can be put in a form suitable for the n-dimensional case by replacing ordinary intervals by n-dimensional intervals. Its graph is a unimodal hypersurface with a single minimum within a specified region. Only in very exceptional circumstances is this the situation in practical optimization problems. In general it is necessary either to find a unimodal region which contains the global minimum, or to distinguish local minima from the global minimum. When the objective function f is to be reduced below a set figure it is often unimportant if a minimum found is local or not.

For convex functions, local minima are also global. The definition of a convex (or concave) function given in Section 1.3 can be applied directly to the n-dimensional case by interpreting the arguments x and y as vectors and the expression $\lambda x + (1 - \lambda) y$ in terms of scalar multiplication and vector addition.

Useful results relating the two concepts are the following (see Karlin (1959), Appendix B). If a twice-differentiable function defined on an n-dimensional interval is unimodal, then its Hessian is positive definite if, and only if, it is strictly convex. The equivalence is actually true without the condition of unimodality, but strictly convex functions are unimodal.

Saddles

Geometrically, a saddle point of a function of two variables looks like a maximum or minimum depending on the direction of a straight line path across the saddle in the plane (cf. Point B in Figure 2). Formally, a point $\bar{\mathbf{x}}$ is a local *saddle-point* of a function f of a vector argument if, perhaps after suitable relabelling of the parameters, $\bar{\mathbf{x}}$ may be written for some $j (1 < j < n)$ as

$$\bar{\mathbf{x}}^{\mathrm{T}} = (\bar{x}_1, \ldots, \bar{x}_j; \bar{x}_{j-1}, \ldots, \bar{x}_n) = (\bar{\mathbf{x}}_1^{\mathrm{T}}, \bar{\mathbf{x}}_2^{\mathrm{T}}) \qquad (1.18)$$

with the property that

$$f(\mathbf{x}_1, \bar{\mathbf{x}}_2) \leqslant f(\bar{\mathbf{x}}_1, \bar{\mathbf{x}}_2) \leqslant f(\bar{\mathbf{x}}_1, \mathbf{x}_2) \qquad (1.19)$$

for all $\mathbf{x}_1$, $\mathbf{x}_2$ in some suitable hyper-interval containing $\bar{\mathbf{x}}_1$, $\bar{\mathbf{x}}_2$. Since for $\bar{\mathbf{x}}_1$ fixed the function is minimized at $\bar{\mathbf{x}}_2$, and for $\bar{\mathbf{x}}_2$ fixed it is maximized at $\bar{\mathbf{x}}_1$, the Jacobian $\mathbf{g}$ of a differentiable function will be zero at a saddle point. As it is unlikely that a saddle point will ever be located precisely by a numerical procedure, an optimization process will normally proceed to one of the nearby minima. Optimization methods utilizing the Jacobian therefore do not suffer unduly from the presence of saddle points.

Narrow valleys

Long curved narrow valleys are especially troublesome to simpler optimization procedures since a small change in the variables can give unpredictably large changes in the value of the objective function. Progress to the optimum is therefore slow, and several special techniques have been developed for use in this situation. The problem then becomes one of recognition of the existence of the valley.

Difficult minima

Figure 3 illustrates two minima which exist in a unique and well defined manner yet are very difficult to treat by classical methods. If the minimum lies on the constraint boundary then even the techniques of constrained optimization may not work. Similarly, minima on the path of a discontinuity in gradient are difficult to locate by conventional methods. Provided the problems can be recognised, special techniques can be applied.

Specification of the objective function

The objective function is normally stored in the computer as a subroutine or procedure which is called by the main optimization program to evaluate f for current values of the variable x. When the function f is known analytically this is usually straightforward. The equation for f is then programmed for computation as efficiently as possible. Functions which are known only numerically are tabulated and stored within the computer in arrays. Since the computer memory requirements for n-dimensional arrays increase as array size to the power of the dimension, this method of specifying the objective function is limited to a few dimensions only. Numerical arrays are nevertheless useful since values for only some variables of x may need to be stored in this way. Evaluation is more complicated in this case, however, since an interpolation procedure is needed to compute E between the tabulated points.

The arguments of f are often widely different in type and magnitude. When such a function is plotted on graph paper effort is taken to make the graph fit the paper and to fully utilise the available space. In a similar way variables must be scaled for the computer so that the problem has symmetry and the effects of the variables are as similar as possible. Experimental and theoretical work has shown that the best choice of scales are those which make the contours as near to spherical as possible. It is therefore important to try to choose units of measurement such that a unit change in any variable will give equal changes in the objective function value $E = f(\mathbf{x})$. If the function is sufficiently known to allow suitable scaling, convergence will be faster and more accurate results obtained.

Exercises 1

1 Plot a graph of the function given by

$$E = |\cos \pi x| - 10(x - \tfrac{1}{4})^2$$

in the region $0 \leqslant x \leqslant 1$.
State whether the function is unimodal or convex in the region 0 to 1.

2 Plot contours of the function given by

$$E = [10 - (x_1 - 1)^2 - (x_2 - 1)^2]^2 + (x_1 - 2)^2 + (x_2 - 4)^2$$

in the region $-4 \leqslant x_1 \leqslant 6$, $-4 \leqslant x_2 \leqslant 6$. Identify maxima, minima, saddle points, and valleys.

3 Repeat question 2 if the function is subject to:

(i) the constraint $x_1 + x_2 - 3 \leqslant 0$,
(ii) the additional constraint $x_2 = -1$.

CHAPTER TWO

Single variable optimization

2.1 Review of methods

The problem considered in this chapter is that of minimizing the value E of f which is a function of one variable only, i.e.

$$\text{minimize } E = f(x).$$

The function is assumed to be unimodal, but not necessarily continuous, within a search interval defined by a lower limit of x at x_L and an upper limit at x_U. Since optimization in one dimension is an essential part of many multidimensional methods, efficiency of the various methods discussed will be closely examined.

Two classes of method are available for single variable optimization.

(i) The function $f(x)$ is approximated by a known function which can be analysed to find a minimum. Usually the approximating function is a polynomial of low-degree. These methods are called *approximation methods*.

(ii) The size of the interval which contains the minimum is reduced by evaluation of the function f at suitable search points. Although several techniques for placing these points are available they are all basically similar. Such techniques are termed *search methods.*

The main difference between the two classes is that approximation methods can only be applied to continuously differentiable functions, while search methods can be applied to any unimodal function.

Approximation methods

The obvious choice for a function which will approximate $f(x)$ in the region of a minimum is either a quadratic or cubic polynomial. The cubic

$$p(x) = a_0 + a_1 x + a_2 x^2 + a_3 x^3 \tag{2.1}$$

can be differentiated to yield

$$\frac{dp}{dx} = a_1 + 2a_2 x + 3a_3 x^2. \tag{2.2}$$

The optimum value of x is then given by one of the roots of equation 2.2 with $dp/dx = 0$, i.e.

$$x_{\min} = \frac{-a_2 \pm \sqrt{a_2^2 - 3a_1 a_3}}{3a_3}. \tag{2.3}$$

The appropriate root for the minimum would normally be apparent, but a check of the second derivative can be made if necessary.

In order to find the cubic approximation, the values of a_0, a_1, a_2, and a_3 must be determined by solving a set of four linear equations derived from 2.1 using four evaluations of the function f. Alternatively, when f is differentiable the values can be obtained from two evaluations of f and two evaluations of the derivative of f. Since the value of x at the minimum of the cubic is only an approximation to the minimum of the given function it may be necessary to repeat the process iteratively using the latest value for $x_{\min}$ in place of one of the four points used in the previous iteration.

If three points can be selected to span the minimum value of f it is usually sufficient to approximate the function with the quadratic

$$p(x) = a_0 + a_1 x + a_2 x^2. \tag{2.4}$$

The approximate position of the minimum of f is then derived as before by setting $dp/dx = 0$, giving therefore,

$$x_{\min} = -\frac{a_1}{2a_2}.$$

Again the process can be repeated iteratively.

When the three points are equally spaced the solution for a_0, a_1, and a_2 is simplified. Let the three points x_1, x_2, and x_3 be equally spaced in such a way that $x_1 = x_2 - s$ and $x_3 = x_2 + s$. The function values at the three points are E_1, E_2, and E_3 respectively. Three simultaneous linear equations are obtained by substitution into equation 2.4 and these may be solved for a_0, a_1, and a_2. The expression for $x_{\min}$ in terms of the function values is then derived as

$$x_{\min} = x_2 + \frac{s(E_1 - E_3)}{2(E_3 - 2E_2 + E_1} \qquad (2.5)$$

The algorithm of Davies, Swann, and Campey (Box *et al*, 1969) utilises a search method to find three equally spaced points which span the minimum. Expression 2.5 is then used to predict the minimum point. This algorithm is described fully in Section 2.4 after search techniques have been discussed.

The main objections to approximation methods for single variable optimization are the following.

(i) Functions which are discontinuous cannot be minimized with accuracy since approximation gives a poor fit to a discontinuous curve.

(ii) Approximation methods do not normally give any estimate of the accuracy to which minima are determined.

(iii) When a minimum lies outside the range of the points used to determine the polynomial coefficients the accuracy of the minimization is poor. This is because the behavior of the approximating curve is unpredictable outside the range of the approximation. For these reasons search methods are often preferred generally. Of the methods involving approximation, the algorithm of Davies, Swann, and Campey is recommended since it overcomes (iii) while cutting down the number of function evaluations necessitated by search methods. An efficient and widely used alternative (Coggins, 1964) is to use an iteration of the Davies, Swann and Campey method to bracket the minimum, followed by successive quadratic approximations from three unequally spaced points provided by the predicted minimum at the previous iteration and the two best of the three points x_1, x_2, and x_3 used to predict it according to the unequally spaced version of 2.5,

$$x_{\min} = \frac{1}{2}\frac{(x_2^2 - x_3^2)E_1 + (x_3^2 - x_1^2)E_2 + (x_1^2 - x_2^2)E_3}{(x_2 - x_3)E_1 + (x_3 - x_1)E_2 + (x_1 - x_2)E_3} .$$

Search methods

Evaluation of a unimodal function f at one point in an interval gives no information on the location of the minimum. An evaluation of the function at a second point, however, allows reduction in the size of the region in which the minimum must be located. The result of two typical function evaluations at search points x_a and x_b is shown in Figure 4.

The search shows that $f(x_b) > f(x_a)$. For a unimodal function this fact may be used to make a prediction regarding the location of the minimum.

Indeed, $x_{\min}$ cannot be located between x_b and x_U. For if it were, then as x is varied from $x_{\min}$ through x_b and x_a toward x_L, the value E must reduce since $f(x_a) < f(x_b)$. This contravenes the definition of unimodality given in Section 1.3 which states that as x is varied from $x_{\min}$ to x_L the value E must increase. Since the function is unimodal it follows that $x_{\min} \ngtr x_b$ and thus $x_L < x_{\min} < x_b$.

If the result of the search had been reversed, so that $f(x_b) < f(x_a)$, then the argument is also reversed and the conclusion $x_a < x_{\min} < x_U$ is reached. If the search had given $f(x_a) = f(x_b)$, then combination of the two cases above shows that $x_a < x_{\min} < x_b$.

When the function f is explicitly given and differentiable, evaluation of its derivative f' at a single point leads to a similar prediction of the location of the minimum. In the case

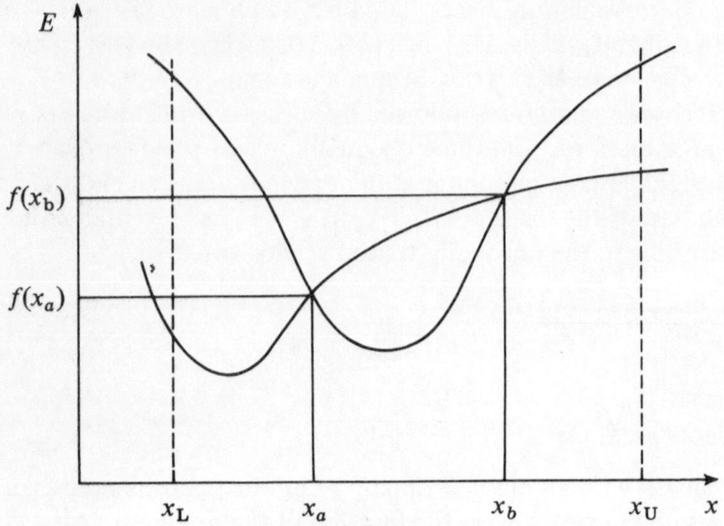

Fig. 4. *Outcome of a two point search.*

considered in Figure 4 at point x_a:

$$\begin{aligned} &\text{if} \quad f'(x_a) > 0, \quad \text{then} \quad x_L < x_{min} < x_a \,, \\ &\text{if} \quad f'(x_a) < 0, \quad \text{then} \quad x_a < x_{min} < x_U \,, \\ \text{and } &\text{if} \quad f'(x_a) = 0, \quad \text{then} \quad x_{min} = x_a \,. \end{aligned}$$

The derivation of this result is straightforward if the first derivative is approximated by $\{f(x_b) - f(x_a)\}/(x_b - x_a)$ as $(x_b - x_a)$ becomes small.

The result can be made the basis of an efficient search routine requiring only derivative evaluations by predicting x_{min} at each iteration by a point x_c using the linear approximation to $f'(x_{min})$ given by

$$f'(x_{min}) - f'(x_a) \approx \left(\frac{x_{min} - x_a}{x_b - x_a} \right) [f'(x_b) - f'(x_a)] \,. \quad (2.6)$$

Treating 2.6 as an equation, using the fact that $f'(x_{min}) = 0$, and substituting x_c for x_{min} yields

$$x_c = x_a - (x_b - x_a)\frac{f'(x_a)}{f'(x_b) - f'(x_a)} \cdot$$

If $f'(x_a)$ and $f'(x_b)$ are of opposite signs, x_c will interpolate x_a and x_b, i.e. $x_a < x_c < x_b$; if they are of the same sign, x_c will lie outside the interval from x_a to x_b (see Figure 4). In either case, f' is evaluated at x_c and to begin the next iteration x_c is used to replace that point x_a or x_b at which f' is greatest. Iterations are terminated when f' is sufficiently close to zero. The resulting search technique is called the *secant method.* After an initial derivative evaluation at two points, it requires only a single derivative evaluation at each iteration, and hence is of use with multidimensional methods requiring gradient information. It has been shown to be both theoretically and practically superior in efficiency to the classical Newton–Raphson technique (for finding a zero of f') which requires evaluation of both the first and second derivatives of f (see, for example, Fox and Mayers (1968), Chapter 4). We shall treat a multidimensional version of the Newton–Raphson method in the next chapter.

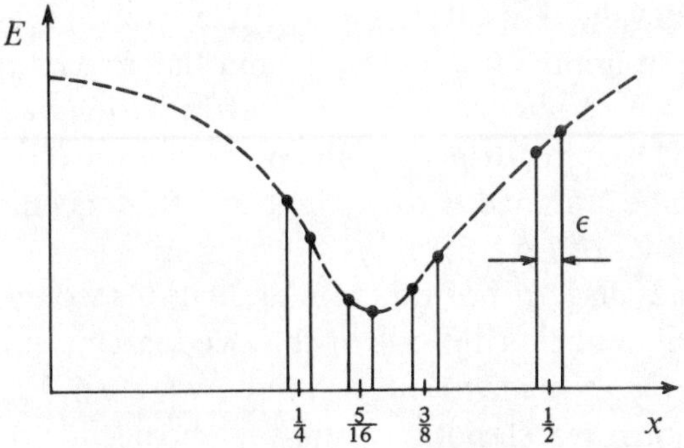

Fig. 5. *Dichotomous search of the interval 0 to 1.*

Optimization in one dimension by search based on function evaluations only is intended to maximize the reduction of the interval known to contain the minimum in the least number of evaluations. The method which is usually the first to come to mind is to successively halve the interval containing the minimum by placing pairs of test points close together at the centre of each interval. This method is illustrated in Figure 5 for an interval of unit length and is called *dichotomous search*.

In this case four pairs of evaluations have reduced the interval from $0 < x_{\mathrm{min}} < 1$ to $5/16 - \epsilon/2 < x_{\mathrm{min}} < 3/8 + \epsilon/2$. The small difference ϵ between the evaluations of each pair is limited either by the data or by numerical errors due to rounding or truncation in computation. It sets a limit of resolution to the optimization. If ϵ is ignored, the reduction of the interval is $(\frac{1}{2})^{n/2}$, where n is the number of function evaluations. For $n = 14$ the reduction is to less than 1% of the initial interval. If $\mathbf{I}_n$ is the interval after n evaluations and $\mathbf{I}_0$ is the initial interval, then the corresponding lengths

$$I_n = (\tfrac{1}{2})^{n/2} I_0.$$

Although the dichotomous search method looks very reasonable, it does seem inefficient to put two search points so close together at each interval reduction. Consider the search interval once again, redrawn in Figure 6, and arbitrarily insert two evaluations at x_a and x_b. The interval $\mathbf{I}_k$, upper and lower search limits $x_{\mathrm{U},k}$ and $x_{\mathrm{L},k}$, and the search points $x_{a,k}$ and $x_{b,k}$, have been given an extra suffix k to indicate the state of the search at the kth interval reduction. The next interval to be searched is $\mathbf{I}_{k+1}$, right or left, depending on the relative values of $f(x_{a,k})$ and $f(x_{b,k})$.

Efficiency in terms of function evaluations would be considerably enhanced if one of the two search points could be used again in the next interval to be searched, $\mathbf{I}_{k+1}$. Only one additional search point would then be needed for each succeeding interval reduction. The basic search method indicates that the minimum lies either in the interval $x_{a,k}$ to $x_{\mathrm{U},k}$, or in the interval $x_{\mathrm{L},k}$ to $x_{b,k}$. Since either interval may contain the minimum they must be equal in length if a known reduction is to be obtained. The two possible intervals, $\mathbf{I}_{k+1}^{\mathrm{R}}$ and $\mathbf{I}_{k+1}^{\mathrm{L}}$, right and left respectively, must then be searched at the next stage of the minimization. If the interval $\mathbf{I}_{k+1}^{\mathrm{R}}$ is assumed to contain the minimum, then a new search point, $x_{b,k+1}$, must be placed in association with $x_{b,k}$ within the interval so that the next possible reduced intervals $\mathbf{I}_{k+2}$ are also equal. These intervals are also marked on Figure 6. It is immediately apparent that the relationship between successive interval lengths is given by

$$I_k = I_{k+1} + I_{k+2}. \tag{2.7}$$

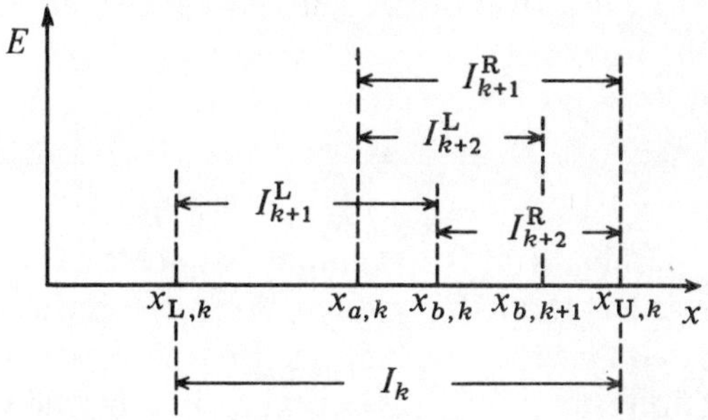

Fig. 6. *Search of interval* $\mathbf{I}_k$. R *right.* L *left.*

The actual lengths of $\mathbf{I}_k$, $\mathbf{I}_{k+1}$, and $\mathbf{I}_{k+2}$ have not so far been determined. These will now be found by considering how the interval reduction must end. If the process is to have a unique predictable overall reduction, the final interval searched must be divided into two halves. This will be accomplished if the pair of search points in the final interval, the last two of n, say, coincide. It is usual to displace one of the final search points by a small amount ϵ determined by the limited resolution of the system. The prediction that the minimum lies in one of the two final equal intervals completes the process.

If a final interval of length I_n is assumed, then equation 2.7 will give values for the length of all previous intervals $\mathbf{I}_1, \mathbf{I}_2, \ldots, \mathbf{I}_{n-1}$. It should be noted that at the final step the interval $\mathbf{I}_{n+1}$ is created equal in length to $\mathbf{I}_n$ but is not used. It is needed, however, to determine the length I_{n-1}, i.e.

$$
\begin{aligned}
I_{n+1} &= I_n \\
I_{n-1} &= I_n + I_{n+1} = 2I_n \\
I_{n-2} &= I_{n-1} + I_n = 3I_n \\
I_{n-3} &= I_{n-2} + I_{n-1} = 5I_n \\
I_{n-4} &= I_{n-3} + I_{n-2} = 8I_n \\
&\text{etc....}
\end{aligned}
$$

The sequence 1,1,2,3,5,8,13,21,34,55,89,144, ... is the well known Fibonacci sequence and search based on this idea is called the *Fibonacci search.* If this sequence of numbers is given by F_0, F_1, F_2, ... then the ratio of the length of the final interval $\mathbf{I}_n$ to that of the initial interval $\mathbf{I}_1$ is given by

$$\frac{I_n}{I_1} = \frac{1}{F_n} . \tag{2.8}$$

After eleven evaluations for example, the interval reduction is 1/144. This is less than 1%, and for comparison the dichotomous search required fourteen evaluations to achieve almost the same reduction. It can be shown that among methods requiring a fixed number of function evaluations Fibonacci search achieves the greatest interval reduction (see, for example, Wilde and Beightler (1967), Chapter 6). Of course this does not mean its efficiency is comparable to methods which utilize derivative information. The secant method is generally practically superior for differentiable functions f.

The main criticism of the Fibonacci search method is that the number of function evaluations must be specified in advance so that the position of the first two evaluations can be determined. This is a nuisance when, for example, f is to be minimized to a known fraction of E_0 and the number of evaluations cannot therefore be predicted.

The placement of the first two evaluations is specified by the requirement that the search should end when the last two evaluations coincide and halve the final interval searched. If this is not assumed to happen, a new criterion for determination of the actual interval lengths in equation 2.7 must be found which does not affect the argument leading to that equation. This can be achieved if the ratio of successive intervals is held constant, i.e.

$$\frac{I_k}{I_{k+1}} = \frac{I_{k+1}}{I_{k+2}} = K \qquad (2.9)$$

so that

$$\frac{I_k}{I_{k+2}} = K^2. \qquad (2.10)$$

Dividing Equation 2.7 by I_{k+2} yields

$$\frac{I_k}{I_{k+2}} = \frac{I_{k+1}}{I_{k+2}} + 1. \qquad (2.11)$$

Substituting from Equations 2.9 and 2.10 gives

$$K^2 = K + 1 \qquad (2.12)$$

so that

$$K = \frac{1 \pm \sqrt{5}}{2}. \qquad (2.13)$$

Taking the positive root, $K = 1.618034$.

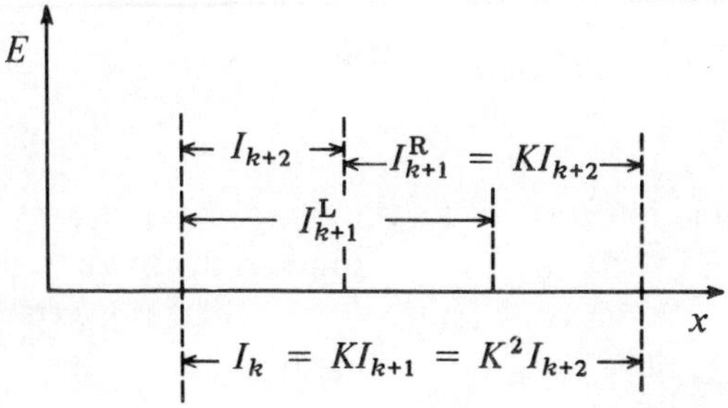

Fig. 7. *Golden section search at the* kth *iteration.* R *right.* L *left.*

A search based on this scheme is called a *golden section*

search because the division of an interval such that the ratio of the smaller to the larger interval is equal to the ratio of the larger to the whole interval is known as the 'golden section'. Successive intervals are shown in Figure 7. An additional feature of the number 1.618034 is that its inverse is 0.618034. This is established by dividing equation 2.12 by K to yield $1/K = K - 1$.

A relationship between the Fibonacci numbers and the constant K can be derived, see for example Wilde (1964), Chapter 2, but it will be sufficient here to state that for a large number of evaluations, say n,

$$F_n \approx \frac{K^{n+1}}{\sqrt{5}}. \tag{2.14}$$

The Fibonacci interval reduction is therefore given by R_F, where

$$R_F = \left(\frac{I_n}{I_1}\right) = \frac{1}{F_n} = \frac{\sqrt{5}}{K^{n+1}}. \tag{2.15}$$

Similarly for the golden section

$$R_{GS} = \left(\frac{I_n}{I_1}\right) = \frac{1}{K^{n-1}}. \tag{2.16}$$

Comparison of the interval reduction can therefore be made, i.e.

$$\frac{R_{GS}}{R_F} = \frac{K^2}{\sqrt{5}} \approx 1.17. \tag{2.17}$$

Thus the final interval for a Fibonacci search is only 17% smaller than for the golden section search for an equal number of function evaluations. This is obtained at the expense of having to specify the number of search points in advance.

2.2. The Fibonacci search

Theory

The theoretical foundation for the Fibonacci search was given in the previous section. It remains to translate the theory into a form suitable for computation and to give an example of its application.

The available data for the search are the upper and lower limits, $x_{U,1}$ and $x_{L,1}$, of the search and the number of function evaluations, n, to give an interval reduction of $1/F_n$. The second subscript on the upper and lower limits indicates the iteration number. At the kth iteration the upper and lower bounds will have been reduced to $x_{U,k}$ and $x_{L,k}$. Also two trial function evaluations at $x_{a,k}$ and $x_{b,k}$ will have been made, giving values of E equal to $E_{a,k}$ and $E_{b,k}$. The size of the search interval $\mathbf{I}_k$ at the kth iteration is equal to $F_{n-k+1}I_n$. The ratio I_k/I_{k+1} is therefore given by the factor F_{n-k+1}/F_{n-k}. The situation is shown in Figure 8.

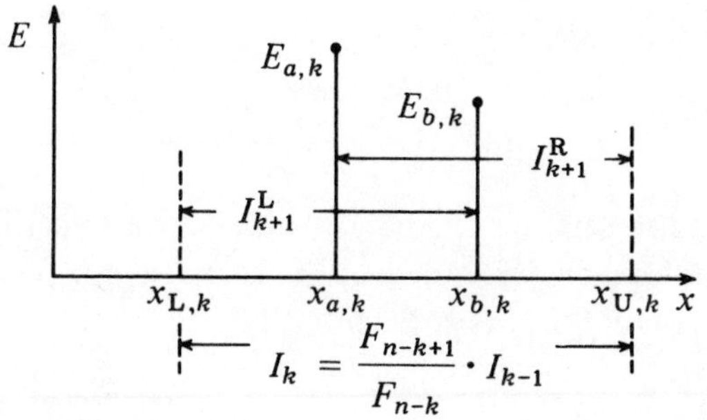

Fig. 8. *Fibonacci search at the* k*th iteration.* R *right.* L *left.*

If $E_{b,k} < E_{a,k}$ the new interval, $x_{L,k+1}$ to $x_{U,k+1}$, to be searched is $\mathbf{I}^R_{k+1}$, which is the right interval from $x_{a,k}$ to $x_{U,k}$, i.e.

$$x_{L,k+1} = x_{a,k} \tag{2.18}$$

$$x_{U,k+1} = x_{U,k}. \tag{2.19}$$

Also the two evaluation points $x_{a,k+1}$ and $x_{b,k+1}$ in the new interval are located respectively at $x_{b,k}$ and at a distance $(F_{n-k-1}/F_{n-1})I_{k+1}$ from $x_{L,k+1}$, i.e.

$$x_{a,k+1} = x_{b,k} \tag{2.20}$$

$$x_{b,k+1} = x_{L,k+1} + \frac{F_{n-k-1}}{F_{n-k}} I_{k+1}. \tag{2.21}$$

Therefore we have

$$E_{a,k+1} = E_{b,k} \tag{2.22}$$

$$E_{b,k+1} = f(x_{b,k+1}). \tag{2.23}$$

If $E_{b,k} > E_{a,k}$ then similar arguments relating to the left interval $\mathbf{I}^{L}_{k+1}$ give

$$x_{L,k+1} = x_{L,k} \tag{2.24}$$

$$x_{U,k+1} = x_{b,k} \tag{2.25}$$

$$x_{a,k+1} = x_{U,k+1} - \frac{F_{n-k-1}}{F_{n-k}} I_{k+1} \tag{2.26}$$

$$x_{b,k+1} = x_{a,k} \tag{2.27}$$

$$E_{a,k+1} = f(x_{a,k+1}) \tag{2.28}$$

$$E_{b,k+1} = E_{a,k}\ . \tag{2.29}$$

By using the relationships 2.18 to 2.29 the six variables can be updated at each iteration and the search continued until x_a and x_b coincide. This is detected not by numerical equality, which is uncertain in a computer, but by counting until $k = n - 2$ giving $F_{n-1} = 2$. At this point the optimization could be terminated with the information that $x_{\min} = x_{a,k+1}$, $E_{\min} = E_{a,k+1}$, and with the error in $x_{\min}$ given by I_1/F_n. The final evaluation however is normally carried out by shifting the final test point in the appropriate direction according to the general search method by a small amount ϵ limited by the resolution of the system and computer. This evaluation halves the interval size but the specific values of x or E are not usually recorded. Then the output information will be that $x_{\min}$ lies within an interval of x equal to I_1/F_n with $E_{\min} \leqslant E_{a,k+1}$. When the interval is to be reduced by a very large factor it is useful to check that numerical inaccuracy has not caused $x_{a,k}$ to exceed $x_{b,k}$. If this occurred the minimization would fail, since the wrong intervals would be assumed to contain the minimum.

The practical search

The practical steps in the computation can now be specified, but a number of minor addenda should be pointed out.

(i) The Fibonacci numbers needed can be obtained, after the first pair, from the difference of the last pair used. Alternatively, they may be stored as the Fibonacci sequence is generated from 2.7 to obtain the first pair.

(ii) Equality of $E_{a,k}$ and $E_{b,k}$ is possible and indicates that the minimum lies between $x_{a,k}$ and $x_{b,k}$. In this case it is not inconsistent to assume that $E_{a,k}$ is less than $E_{b,k}$.

(iii) The test for the end of optimization in the description below is changed from $k = n - 2$ to $k = n - 1$ because of the location of the step which increments k by 1.

Fibonacci search

Step: 1. Input data $x_{L,1}, x_{U,1}, n, f$.

2. Compute the Fibonacci numbers F_n, F_{n-1}.

3. Compute the first two test positions from

$$x_{a,1} = x_{U,1} - \frac{F_{n-1}}{F_n}(x_{U,1} - x_{L,1})$$

$$x_{b,1} = x_{L,1} + \frac{F_{n-1}}{F_n}(x_{U,1} - x_{L,1}).$$

Set $k = 1$.

4. Evaluate the function f at the two test points,

$$E_{a,k} = f(x_{a,k})$$

$$E_{b,k} = f(x_{b,k}).$$

5. Test for the interval which contains the minimum.

If $E_{a,k} \leqslant E_{b,k}$ go to Step 6.

If $E_{a,k} > E_{b,k}$ go to Step 7.

6. Use equations 2.24 to 2.29 to update all information for the next iteration.
   Check that $x_{a,k} < x_{b,k}$.
   Go to Step 8.
7. Use equations 2.18 to 2.23 to update all information for the next iteration.
   Check that $x_{a,k} < x_{b,k}$.
8. Test for the end of the optimization.
   Set $k = k + 1$.
   If $k = n - 1$ go to Step 9.
   Otherwise go to Step 4.
9. Make the final test at $x = x_{a,k} + \epsilon$ and determine the final limits on the interval containing the minimum.
   Output data $x_{L,n}$, $x_{U,n}$,
   $x_{min} = \frac{1}{2}(x_{U,n} + x_{L,n})$,
   $E_{min} \leqslant E_{a,n-1}$.
   If the final test is omitted then
   $x_{min} = x_{a,n-1} \pm (x_{U,1} - x_{L,1})/F_n$.

Example 1. Minimize the function given by $f(x) = x^2$ by Fibonacci search with $x_{L,1} = -5$, $x_{U,1} = 15$, and $n = 7$.

Notes: (i) All results were computed with a slide rule.
(ii) If $E_{b,k} < E_{a,k}$, the new search interval is the right, '*R*', interval and if $E_{b,k} \geqslant E_{a,k}$, the new search interval is the left, '*L*', interval.

The computations are carried out systematically in Table 1 below. Arrows indicate the transfer of existing data from one iteration to the next.

Table 1. Fibonacci search

k	$\frac{F_{n-k-1}}{F_{n-k}}$	$x_{L,k}$	$x_{U,k}$	$x_{a,k}$	$x_{b,k}$	$E_{a,k}$	$E_{b,k}$	
1	13/21	−5.0	15.0	2.62	7.38	6.88	54.6	L
2	8/13	−5.0	7.38	−0.24	2.62	0.058	6.88	L
3	5/8	−5.0	2.62	−2.14	−0.24	4.67	0.058	R
4	3/5	−2.14	2.62	−0.24	0.72	0.058	0.52	L
5	2/3	−2.14	0.72	−1.2	−0.24	1.44	0.058	R
6	1/2	−1.2	0.72	−0.24	−0.24 +0.01	0.058	0.053	R
7	1	−0.24	0.72					

Answer Either, $x_{min} = -0.24 \pm 20/F_7$, $E_{min} = 0.058$, or, if the final test is carried out at $x_{b,k} = -0.24 + 0.01$, $-0.24 < x_{min} < 0.72$, and $E_{min} \leqslant 0.053$.

2.3. The Golden Section Search

Theory

The search by golden section differs from the Fibonacci search in two ways.

(i) In place of the ratio F_{n-k-1}/F_{n-k}, which appears in equations 2.21 to 2.26, the fixed ratio $1/K = K - 1 = 0.618034$ is substituted. The necessity to know n is removed. Equations 2.21 and 2.26 are replaced respectively by the equations

$$x_{b,k+1} = x_{L,k+1} + KI_{k+1} \tag{2.30}$$

$$x_{a,k+1} = x_{U,k+1} - KI_{k+1}, \tag{2.31}$$

with K given by $(\sqrt{5} + 1)/2$.

(ii) Since n is no longer available a new test must be established to end the search. Only two variables, E and x, are available. The simplest test to locate $x_{\min}$ requires the interval I_k containing $x_{\min}$ to be less than a specified ϵ.

The practical search

The practical steps in the golden section search are similar to those for the Fibonacci search but for clarity they are given in full.

Golden section search

Step: 1. Input data $x_{L,1}, x_{U,1}, \epsilon, f$.

2. Compute the first two test points,

$$x_{a,1} = x_{U,1} - K(x_{U,1} - x_{L,1})$$

$$x_{b,1} = x_{L,1} + K(x_{U,1} - x_{L,1}).$$

Set $k = 1$.

3. Evaluate the function f at the two test points,

$$E_{a,k} = f(x_{a,k})$$

$$E_{b,k} = f(x_{b,k}).$$

4. Test for the interval which contains the minimum.

If $E_{a,k} \leqslant E_{b,k}$ go to Step 5.

If $E_{a,k} > E_{b,k}$ go to Step 6.

5. Use equations 2.24, 2.25, 2.31, and 2.27 to 2.29 to update all information for the next iteration.
Check that $x_{a,k} < x_{b,k}$.
Go to Step 7.

6. Use equations 2.18 to 2.20, 2.30, 2.22, and 2.23 to update all information for the next iteration.
Check that $x_{a,k} < x_{b,k}$.

7. Test for the end of the optimization.
If $I_k < \epsilon$ go to Step 8.
Otherwise set $k = k + 1$. Go to Step 4.

8. Output data $x_{L,n}, x_{U,n}$ and either $E_{min} < E_{a,n}$ (right interval), or $E_{min} < E_{b,n}$ (left interval) (n is the final iteration number).

Example 2. Minimize the function given by $f(x) = x^2$ by golden section search with $x_{L,1} = -5$, $x_{U,1} = 15$, and $\epsilon = 1.5$.

Notes: (i) All results were computed by slide rule.
(ii) Right, R, and left, L, intervals are as in Example 1.

The computations are carried out systematically in Table 2 below. The transfer of existing information from one iteration to the next is indicated by arrows.

Table 2. Golden section search

k	$x_{L,k}$	$x_{U,k}$	$x_{a,k}$	$x_{b,k}$	$E_{a,k}$	$E_{b,k}$	
1	−5.0	15.0	2.64	7.36	6.96	54.1	*L*
2	−5.0	7.36	−0.29	2.64	0.084	6.96	*L*
3	−5.0	2.64	−2.08	−0.29	4.33	0.084	*R*
4	−2.08	2.64	−0.29	0.84	0.084	0.71	*L*
5	−2.08	0.84	−0.96	−0.29	0.92	0.084	*R*
6	−0.96	0.84	−0.29	0.15	0.084	0.023	*R*
7	−0.29	0.84			0.023		

After the same number of function evaluations as in Example 1, the following result is available,

$$x_{L,7} = -0.29, x_{U,7} = 0.84, E_{min} \leqslant 0.023$$

In comparison, the final interval containing x_{min} gives a ratio

$$\frac{x_{U,7} - x_{L,7} \text{ (for golden section)}}{x_{U,7} - x_{L,7} \text{ (for Fibonacci)}} = \frac{1.13}{0.96} = 1.18.$$

This is close to the theoretical value of 1.17 obtained in equation 2.17.

2.4. The Algorithm of Davies, Swann, and Campey

Theory

The algorithm of Davies, Swann, and Campey is a combination of a search method which produces three equally spaced x points that span the minimum and the application of expression 2.5. This gives the minimum of a quadratic fitted to the function at the three points. The theory of the quadratic approximation was given in Section 2.1 and it remains to describe the search method used.

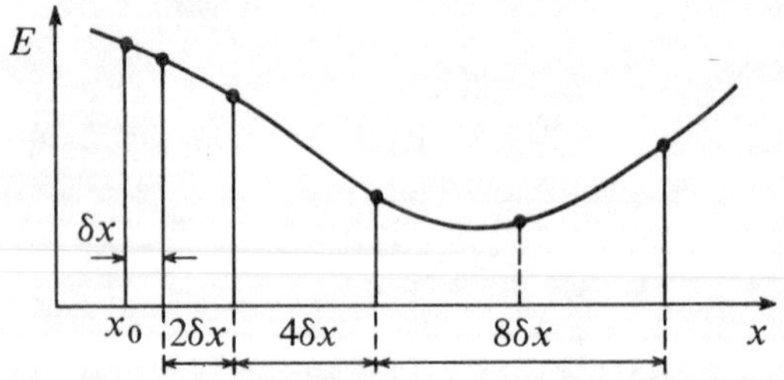

Fig. 9. *Davies, Swann, and Campey's search.*

The basic technique of the search is illustrated geometrically in Figure 9. Search of the function $f(x)$ commences at x_0. An initial increment δx is either specified or chosen. In the successful direction of search, in this case the positive direction of x, successively doubled increments (i.e. δx, $2\delta x$, $4\delta x$, $8\delta x$,..., etc.) are added to x and the function evaluated at each point. This process is stopped when the function shows an increase in value compared to the previous evaluation. Provided the function is unimodal, the minimum must lie within the interval spanned by the last three evaluated points. In Figure 9 the minimum must therefore lie between the points

$$x_0 + 3\delta x \text{ and } x_0 + 15\delta x.$$

An additional function evaluation at the centre of the final interval produces four equally spaced evaluations which span the minimum. One of these is redundant for the purposes of quadratic approximation. Therefore, the outermost point, i.e. the point most remote from the evaluation giving a minimum value of the function, is discarded. In Figure 9 the additional evaluation is placed at $x_0 + 11\delta x$. The evaluation at $x_0 + 11\delta x$ gives the least value of the function so the evaluation at $x_0 + 3\delta x$ is discarded. The remaining three evaluations at $x_0 + 7\delta x$, $x_0 + 11\delta x$, and $x_0 + 15\delta x$ span the minimum and are equally spaced $4\delta x$ apart. Expression 2.5 is then used to predict an approximation to $x_{\min}$.

When the evaluation at $x_0 + \delta x$ gives a greater function value than at x_0, the function is evaluated in the negative search direction at $x_0 - \delta x$. If this test also fails to reduce the function value, the three points $x_0 - \delta x$, x_0, and $x_0 + \delta x$ span the minimum as required. If the test at $x_0 - \delta x$ successfully reduces the function value, the negative direction is searched in the same manner as that described for the positive direction.

The process is made iterative by using the predicted minimum from expression 2.5 for the start of the next iteration while reducing δx by a suitable factor K.

Algebraically the search points may be expressed by the following sequence,

$$\begin{aligned}
&x_{0,k} \\
&x_{1,k} = x_{0,k} + \delta x_k \\
&x_{2,k} = x_{1,k} + 2\delta x_k \\
&x_{3,k} = x_{2,k} + 4\delta x_k \\
&\vdots \\
&x_{n,k} = x_{n-1,k} + 2^{n-1}\,\delta x_k\,,
\end{aligned} \tag{2.32}$$

where k has been introduced to indicate the iteration number. The corresponding function values are $E_{0,k}, E_{1,k}, E_{2,k}, \ldots, E_{n,k}$. The search is stopped at the nth function evaluation when $E_{n,k} > E_{n-1,k}$. The increment δx_k is negative for the negative search direction.

If the final evaluation is performed at the point $x_{n,k}$ then an additional function evaluation, $f(x_{e,k})$ is made at the point $x_{e,k} = x_{n,k} - 2^{n-2}\,\delta x_k$. The four function evaluations at the points $x_{n-2,k}, x_{n-1,k}, x_{e,k}$ and $x_{n,k}$ are equally spaced apart by $2^{n-2}\,\delta x_k$. Either the evaluation at $x_{n,k}$ or at $x_{n-2,k}$ is discarded, so that (for n greater than 1) substitution into expression 2.5 yields

$$x_{0,k+1} = x_{n-1,k} + \frac{2^{n-2}\,\delta x_k\,(E_{n-1,k} - E_{e,k})}{2(E_{e,k} - 2E_{n-1,k} + E_{n-2,k})} \qquad (2.33)$$

when $E_{e,k} > E_{n-1,k}$,
and

$$x_{0,k+1} = x_{e,k} + \frac{2^{n-2}\,\delta x_k\,(E_{n-2,k} - E_{n,k})}{2(E_{n,k} - 2E_{e,k} + E_{n-1,k})} \qquad (2.34)$$

when $E_{e,k} < E_{n-1,k}$.

The practical search

The computation can now be put into step by step form. The following points should also be noted.

(i) The sign of δx_k has been incorporated into the factor p which is set to $+1$ or -1 during the computation. The value $p\delta x_k$ must then be substituted into equations 2.33 and 2.34.

(ii) Prior to p being set, $E_{-1,k}$ and $x_{-1,k}$ refer to the evaluation at $x_{0,k} - \delta x_k$. Once p is set, the negative suffix to f and x is no longer necessary.

(iii) The minimization is terminated when the interval between the three points used to fit the quadratic is reduced below a specified value ϵ.

Davies, Swann, and Campey's algorithm

Step: 1. Input data $x_{0,1}$, δx_1, ϵ, K, f. Set $k = 0$.

2. Set $k = k + 1$.
Evaluate f at $x_{0,k}$ and $x_{1,k}$ to give $E_{0,k} = f(x_{0,k})$ and $E_{1,k} = f(x_{1,k})$.

3. Test for positive search direction.
If $E_{1,k} < E_{0,k}$, set $p = +1$ and go to Step 5.
Otherwise go to Step 4.

4. Test for negative search direction.
Evaluate f at $x_{-1,k}$.
If $E_{-1,k} < E_{0,k}$, set $p = -1$ and go to Step 5.
Otherwise go to Step 8.

5. Search until the minimum is spanned.
Either search the positive direction (for $p = +1$), or search the negative direction (for $p = -1$), until $E_{n,k} > E_{n-1,k}$.

6. Insert the extra function evaluation at the centre of the final interval.
Evaluate f at $x_{n,k} - 2^{n-2} p \delta x_k$.

7. Determine the minimum of the fitted quadratic.
If $E_{e,k} > E_{n-1,k}$ use equation 2.33 to give $x_{0,k+1}$.
Otherwise use equation 2.34.
If $2^{n-2}\delta x_k \leqslant \epsilon$ go to Step 9.
Otherwise, set $\delta x_{k+1} = K\delta x_k$ and go to Step 2.

8. Determine the minimum of the quadratic fitted to $x_{-1,k}$, $x_{0,k}$, and $x_{1,k}$.

$$\text{Set } x_{0,k+1} = x_{0,k} + \frac{\delta x_k (E_{-1,k} - E_{1,k}}{2(E_{1,k} - 2E_{0,k} + E_{-1,k})}$$

If $\delta x_k \leqslant \epsilon$ go to Step 9.
Otherwise, set $\delta x_{k+1} = K\delta x_k$, and go to Step 2.

9. Output data, either δx_k or $2^{n-2}\delta x_k$,

$$x_{min} = x_{0,k+1},$$

$$E_{min} = E_{0,k+1}$$

Example 3. Minimize the function given by $f(x) = x^2$ using Davies, Swann, and Campey's algorithm with $x_0 = 15$, $\delta x = 2$, $K = 0.5$, and $\epsilon = 1.0$.

Table 3. Davies, Swann and Campey algorithm

Iteration 1 $x_{0,1} = 15, \quad \delta x_1 = 2.$

Positive search direction

	$x_{0,1}$	$x_{1,1}$
x	15	17
$E = f(x)$	225	289

Negative search direction (Since $E_{1,1} > E_{0,1}$)

	$x_{0,1}$	$x_{1,1}$	$x_{2,1}$	$x_{3,1}$	$x_{4,1}$
x	15	13	9	1	−15
E	225	169	81	1	225

Since $E_{4,1} > E_{3,1}$, stop the search with $n = 4$.

Additional evaluation

	$x_{e,1}$
x	−7
E	49

Since $E_{e,1} > E_{3,1}$, discard the evaluation at $x_{4,1}$.
Use $E_{2,1}, E_{3,1}$, and $E_{e,1}$ in equation 2.33 to give $x_{0,2} = 0$.
Also $2^{n-2}\delta x_1 = 8 > \epsilon$, and $E_{0,2} = 0$.

Iteration 2 $x_{0,2} = 0, \quad \delta x_2 = 1.$

Positive search direction

	$x_{0,2}$	$x_{1,2}$
x	0	1
E	0	1

Negative search direction (Since $E_{1,2} > E_{0,2}$)

	$x_{0,2}$	$x_{-1,2}$
x	0	1
E	0	1

Since $E_{-1,2} > E_{0,2}$, stop the search and use the points $x_{-1,2}, x_{0,2}$, and $x_{1,2}$ to find $x_{0,3}$ from Step 4 giving $x_{0,3} = 0$. Also $\delta x_2 = 1 = \epsilon$, and $E_{0,3} = 0$.

Since $\delta x_2 = \epsilon$ the minimization is terminated.
Output data, $\delta x_2 = 1$, $x_{\min} = 0$, and $E_{\min} = 0$.

The second iteration exhibits ideal behaviour by predicting the minimum at the same point as the first iteration. Since the function is known to be quadratic, the algorithm should converge in one iteration. This is confirmed.

Exercises 2

1 Approximate the function defined by the table of values by a quadratic polynomial and determine the minimum

value. Compare with the actual minimum of the function given by $E = |\cos x|$ over the interpolated interval.

E	$\frac{1}{2}$	0	1
x	60	90	180

2 With reference to Figure 4.
If a search of the unimodal function given by $E = f(x)$ at points x_a and x_b shows that $f(x_a) = f(x_b)$ prove that $x_a < x_{\min} < x_b$.

3 Write out the calculations of the secant method in a step by step form suitable for computation. Specify termination criteria and any practical precautions necessary for using the algorithm on functions which are not unimodal on the search interval.

4 Minimize the function given by $E = |\cos x|$ between $x_{L,1} = 36°$ and $x_{U,1} = 120°$ using dichotomous search for a final interval reduction of at least 1/16 and $\epsilon = 1°$.

5 Repeat Exercise 4 using Fibonacci search.

6 Repeat Exercise 4 using golden section search.

7 Repeat Exercise 4 using the secant method and the results of Exercise 3.

8 Minimize the function given by $E = \cos x$ using Davies, Swann, and Campey's algorithm with $x_0 = 60°$, $\delta x = 10°$, $K = 0.5$, and $\epsilon = 1°$.

9 If access to a computer system is available, write computer subroutines for the secant method, Fibonacci search, golden section search, Davies, Swann, and Campey's algorithm, and the method of successive quadratic approximation with the minimum bracketed using their algorithm. Repeat Examples 1, 2, and 3, and Exercises 5, 6, and 7 to check both the programs and the exercises.

CHAPTER THREE

Multi-variable optimization

3.1 Introduction

Some of the problems encountered in the minimization of functions of n variables were introduced in Chapter 1. One additional problem, best illustrated by an example, is the enormity of hyperspace. Consider a 10 dimensional unit hypercube in which a search has determined that the volume from the origin to the point $\frac{1}{2}$ in each dimension does not contain the minimum. The volume of the $\frac{1}{2}$ unit cube eliminated from the search is equal to $(\frac{1}{2})^{10} \approx 1/1000$. The portion of the unit hypercube which remains is therefore 99.9% of the whole. Any form of complete search is clearly out of the question and attention must be given to the local area of a first estimate to the solution. All methods can therefore guarantee only to find the local minimum.

This chapter will be illustrated by examples of two dimensional searches since contours of functions of two variables can be drawn on the page. Progress toward the minimum can then be clearly plotted and a better understanding of the method should result. All methods are of course applicable to any number of dimensions.

Methods for multi-variable optimization fall naturally into two classes. Although these classes are not completely separate, methods can be divided into *search* methods which use function evaluation only, and *gradient* methods which in addition require gradient information in the form of the Jacobian gradient vector **g**. The so-called *second order* methods, which require the Hessian matrix **H** as well, are included in the latter category.

3.2 Search methods

Search methods for optimization attempt to reduce the value E of the objective function f by the use of tests near to an estimate of the solution. The tests determine a direction of search in which the minimum is expected to lie. The minimum is then either found by a single variable search or *linear search* in the direction determined, or approached by taking a fixed step towards it. Since the direction determined is not necessarily correct, the process is iterative. After each minimization step further searches are carried out until a criterion which defines when the desired minimum has been found is satisfied.

The first method of optimization which comes into the mind of a scientist or engineer in a practical situation is to adjust all the parameters for minimum output in turn, one at a time. This is repeated until no further improvement can be obtained. It will probably also be appreciated by such persons that the results of such adjustments can be extremely complicated if the parameters interact strongly in their effect on the value E of the objective function f.

In Figure 10 the progress of this type of adjustment, called *one at a time search*, is traced for a two dimensional function with elliptical contours. Each variable is adjusted with all the others fixed, so that each step of one overall iteration consists of a linear search parallel to each one of the co-ordinate axes. A zig-zag path is therefore taken on the way to the minimum. Progress tends to be somewhat slow because the direction of movement is rarely in the direction of the minimum. On long narrow valleys for example, an excessive number of very small steps are taken. Since each step involves a linear search, each with a large number of function evaluations, the one at a time

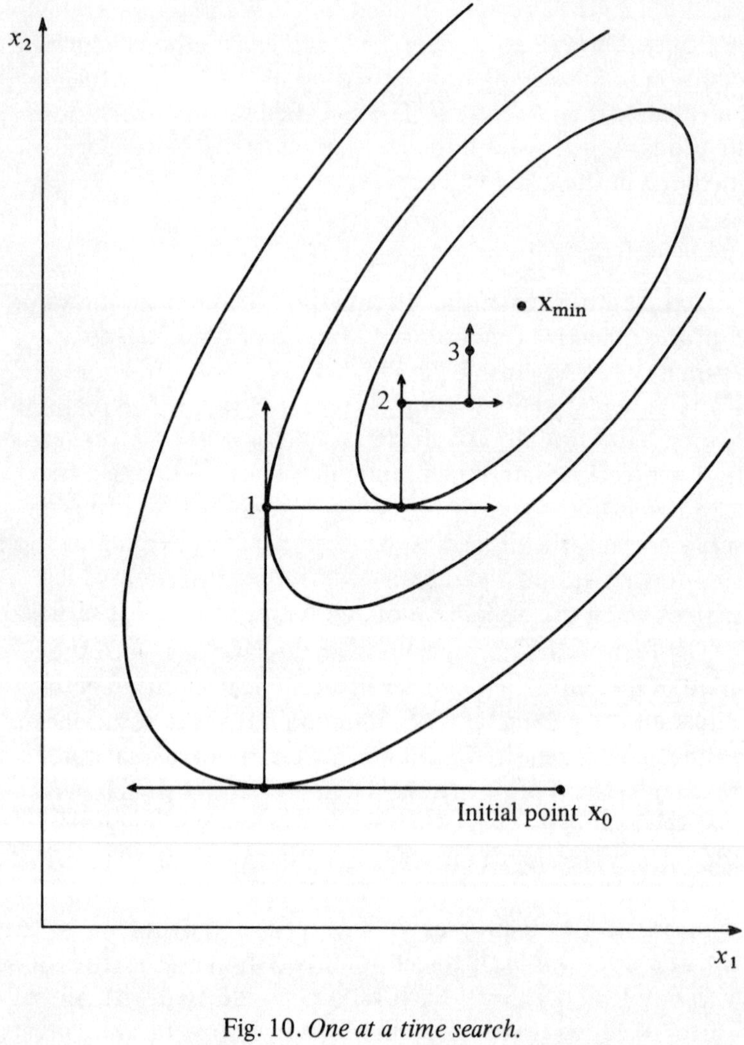

Fig. 10. *One at a time search.*

search must be considered of limited use in situations involving other than very simple functions.

The path of successive minima at the end of each iteration, which are marked 1, 2, and 3 in Figure 10, suggests that it would be very much more efficient to search the direction 1 to 2 rather than the direction parallel to the x_1 axis. The line connecting minima at the beginning and end of each iteration appears to point toward the minimum. This observation leads to the development of the idea behind several optimization techniques which employ a local search to define a direction of movement. In the interest of speed and simplicity a fixed step is taken in the defined direction instead of performing a linear search. Both 'simplex' and 'pattern' search utilise this approach.

The simplex search technique

The *simplex search* technique determines the direction of movement in a two dimensional search problem using only three observations. These observations are placed at the vertices of an equilateral triangle and the function f is evaluated at each point. The direction of movement is simply that from the vertex with the maximum value of E through a point midway between the remaining two vertices. The distance moved is such that the new search position also forms an equilateral triangle with the common base. A local exploration and move of this type is shown in Figure 11. In this example the initial triangle consists of the points 1, 2, and 3 marked 'o'. Vertex 1 gives the largest value E of the function and is therefore reflected about the remaining vertices 2 and 3 to give a new point 4 marked 'x' which forms the equilateral triangle 2, 3, 4.

The generalization of this idea to n dimensions is carried out by selection of $n + 1$ mutually equidistant points which define a simplex (an n dimensional version of the tetrahedron in space). A new simplex is generated by reflection of the vertex which gives the maximum value of E through the remaining n points.

The input information for a simplex search is the same as that required for all search methods of optimization. The function given by $E = f(\mathbf{x})$ and a first approximation $\mathbf{x}_0$ must of course be specified. Search techniques which take steps toward the minimum usually require in addition that the step size be specified. As the minimum of the function is approached the step size is reduced to improve the resolution of the search.

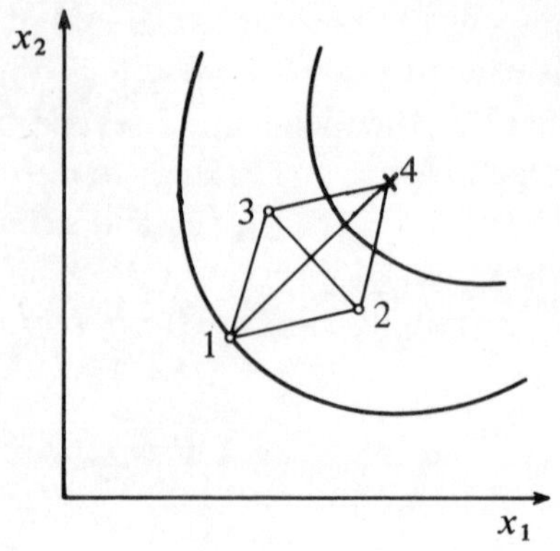

Fig. 11. *Local exploration for the simplex search technique.* o *initial points.* x *new point.*

The search can then be terminated when the step size falls below a specified minimum value.

For simplex search the step size is replaced by a single value which gives the length of one side of the simplex. The use of equal units in each dimension points to the importance of correct scaling for this search method. A lower limit to the length of a simplex side is also required so that the search can be terminated.

The progress of the two dimensional search shown in Figure 12 will be used to illustrate the search procedure and to demonstrate the action taken when the search oscillates. It is then natural to introduce a contraction of the size of the simplex in order to overcome one of the two modes of oscillation.

The first simplex in Figure 12, an equilateral triangle in two dimensions, is defined by the vertices numbered 1, 2, and 3. Vertex 1 represents the first approximation to the minimum. The equilateral triangle 1, 2, 3 is constructed with its side set equal to a specified length. The contours show in this case that reflection is required about the line that joins vertices 2 and 3. Point 4 is then generated and the function evaluated at that point. Since the contours show that the next predicted point

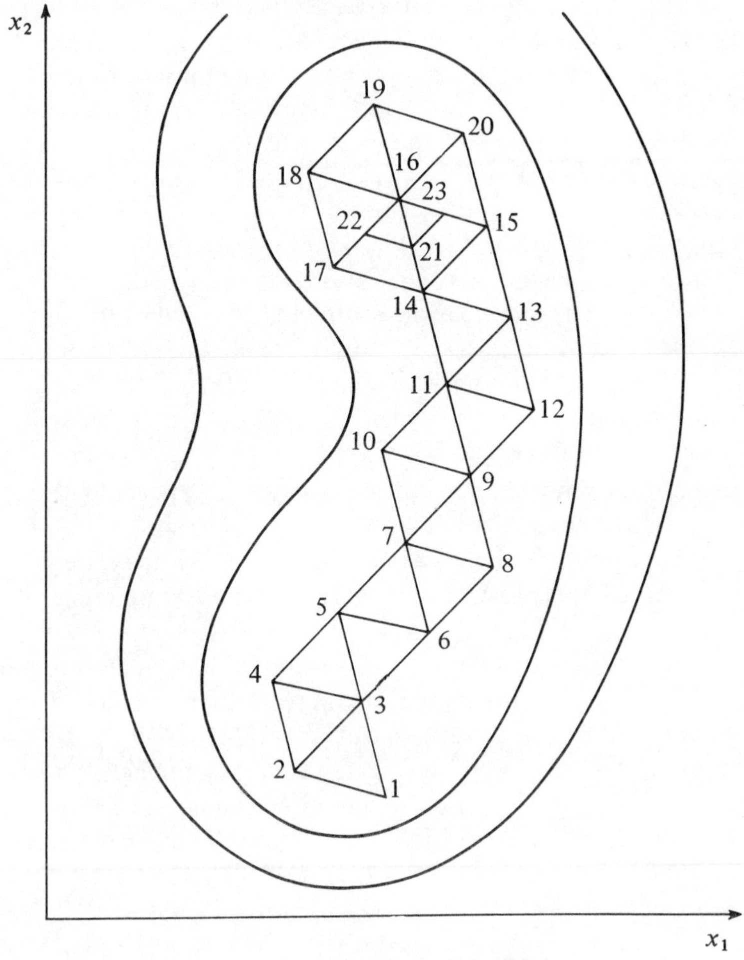

Fig. 12. *The simplex search technique.*

is back at vertex 1, the first mode of oscillation has occurred. This situation can be recognised as occurring when the most recently introduced point gives a maximum value of the function and is therefore rejected.

The solution in this situation is quite straightforward. Instead of rejecting vertex 4, the vertex with the second largest function value is rejected. In this case, therefore, vertex 2 is rejected and reflection takes place about the vertices 3 and 4. The search then progresses down the valley until at vertex 16 a second type of oscillation occurs. In this case the triangles rotate about one vertex until eventually the search returns to the triangle with vertices 14, 15, and 16. In n dimensions exact correspondence does not occur due to machine rounding error. Detection of this oscillation must therefore rely on the fact that one vertex of a simplex is used repeatedly.

As soon as this oscillation is detected, the size of the simplex is contracted by a factor of $\frac{1}{2}$ while the vertex on which the rotation is based is retained. In Figure 12, therefore, a new triangle with vertices 16, 21, and 22 is generated and tested. The next vertex, 23, is very close to minimum and several successive rotations and contractions can be expected before the search is terminated when the simplex has shrunk below the size specified. Rotational oscillations of this type are characteristic either when the search is close to minimum, or when a narrow valley is nearby. In either case a contraction in size is necessary to resolve the direction of movement required.

Several improvements to simplex search are available. The most useful of these is due to Nelder and Mead (1965). It incorporates expansion as well as contraction of the simplex so that movement toward the minimum can be accelerated. Also the simplex search can be arranged to incorporate some form of linear search in the direction of the reflected vertex. Some experiments of Parkinson and Hutchinson (1972) suggest that the simplex method with Nelder and Mead's modifications is very effective in finding the minimum and roughly comparable in efficiency to more advanced methods for problems in many dimensions, say $n > 100$. Effective use of the basic simplex method described is limited to well-behaved functions in only a few dimensions, however, and the algorithm will not be developed further here.

The pattern search

The basic *pattern search* takes incremental steps after suitable directions have been found by local exploration. If the search progresses well in terms of decrementing the objective function, the step size is increased. If it is not progressing, either because the minimum is near or because of difficulties (e.g. a narrow valley), the step size is reduced. When the step size is reduced below a set figure the search is ended.

If a two dimensional search is considered, the following data is available.

(i) A function given by $E = f(x_1, x_2)$ which is to be minimized.

(ii) A first approximation $x_{1,0}, x_{2,0}$.

(iii) Initial parameter increments $\delta x_1, \delta x_2$.

(iv) Parameter increments below which the search is terminated.

The second suffix to the parameters x_1, x_2 is used to signify the iteration number. At each iteration the value of x_1 and x_2 is used as a *base* for local explorations. When a direction has been found a *pattern move* is made to a new base.

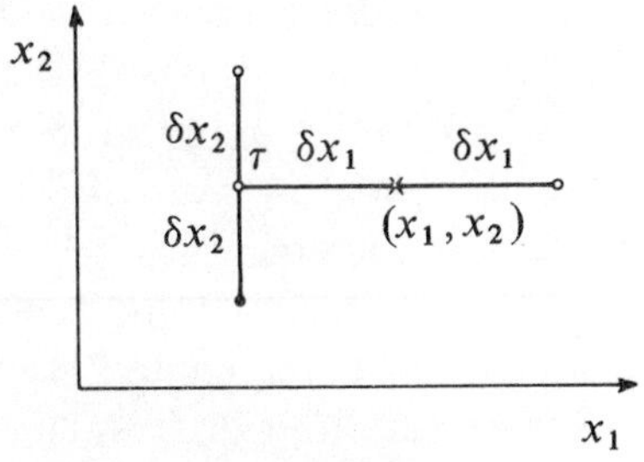

Fig. 13. *Local exploration for the pattern search.*

The situation in the two dimensional case is shown in Figure 13. Local exploration at each base is made by evaluating f at points on either side of the base shifted along each axis in turn by an amount equal to the increment in the current direction. Instead of the simple cross pattern which would result from retaining the original base, however, as each axis is examined a temporary base is set up at the minimum value so far determined. In Figure 13 the original base point is marked 'x'. Evaluation of the function is made at the base and at each point marked ○ or ●. The least value of the function during the exploration has been determined at the point marked ●. It is possible that the base point itself may be found to be the minimum in one, several, or all directions of search. The point marked 'τ' in Figure 13 has become the temporary base for the exploration in the x_2 direction in this example since it gave the minimum value of the function f at the three points evaluated in the x_1 direction. Search in the x_2 direction is therefore shifted to the points $(x_1 - \delta x_1, x_2 \pm \delta x_2)$.

The action taken once the local exploration has been carried out depends on the sequence of moves already made as part of the current pattern. Each pattern consists of a series of successful moves, in each of which the value E is reduced. A pattern is eventually terminated by failure to reduce E. A new pattern is started from the last minimum point determined. In order to demonstrate various modes of action, the progress of a single pattern will be followed with reference to Figure 14.

At the first base point of a pattern, $(x_{1,0}, x_{2,0})$, local exploration as shown in Figure 13 should produce a nearby point at which the function value is less than at the base. The direction from the base x to this point ● is used as the direction of movement. If the base point is found also to be the minimum point, then the parameter increments are all halved and a new exploration made. This process can be repeated until either a direction is determined, or the increments are smaller than the specified limits below which the search is terminated. In Figure 14 the pattern move is made from base point 1, $(x_{1,0}, x_{2,0})$, to base point 2, $(x_{1,1}, x_{2,1})$, by extension of the line from x to ● to twice its length. Increments of $2\delta x_1$ and $2\delta x_2$ have therefore been added in the x_1 and x_2 directions respectively. At base 2, local exploration establishes a new minimum at the point $(x_{1,1} + \delta x_1, x_{2,1})$. The direction of search must now veer off to the right. This is automatically achieved by the method of pattern movement.

Subsequent to the first move of the pattern, pattern moves are made in the direction determined by a line from the

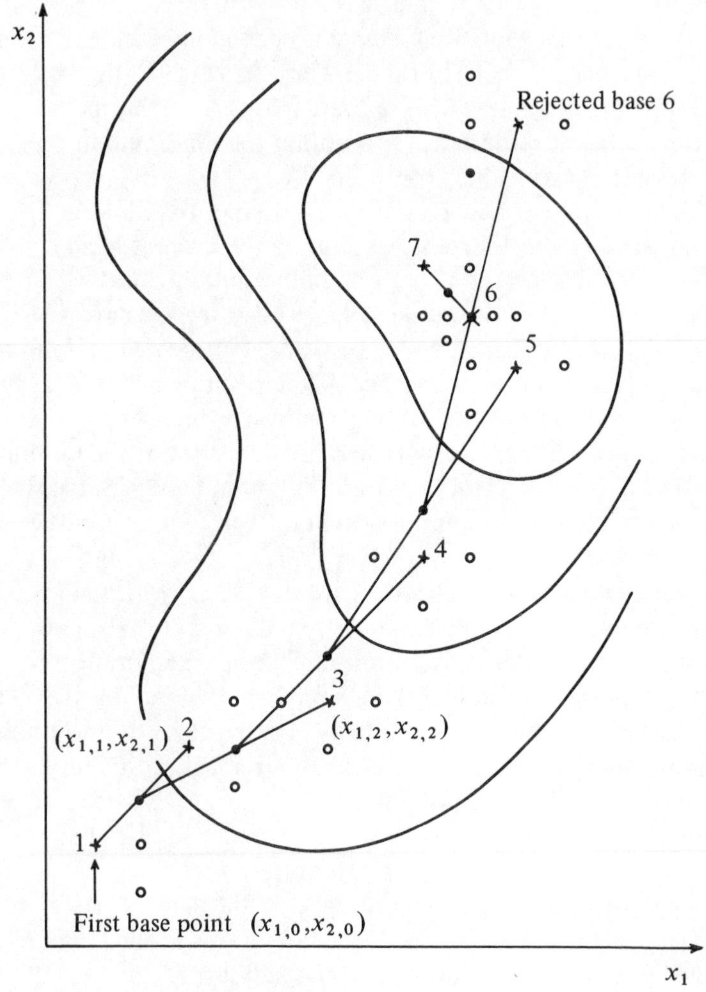

Fig. 14. *The pattern search.*

minimum point of the previous local exploration through the minimum of the current exploration. A new base is established at a distance from the second point equal to that between the two minima. The sequence of moves up to base 5 illustrates the action of this rule. Note that it increases the size of the movement to the next base as a successful direction is established. The actual movement made along each axis from the current minimum to the next base is equal to the algebraic sum of the increments which were added to each previous base in the pattern in order to obtain the minimum in the corresponding local exploration.

The move to a new base is not carried out if the function evaluations at the current base and in the current local exploration are greater than the minimum previously determined. The contours in Figure 14 show a turn in the valley in the region of base 5 so that the projected region of base 6 is well out of the valley. The move through the base 6 minimum found from local exploration around base 6, is therefore rejected. A new base 6 is established at the minimum found during the local exploration around base 5. Since the projected move was a failure, the new base 6 becomes the first base of a new pattern.

The first set of explorations around base 6 with the initial increments of δx_1 and δx_2 fail to find a direction because the process is now close to an optimum. A new exploration is therefore made with increments reduced by a factor of $\frac{1}{2}$. The final base, labelled 7 on Figure 14, is very close to optimum. Improvement will only be made by several successive reductions in the increment size until eventually the step size is reduced below the specified limit and the optimum is located as accurately as this specification requires.

We are now in a position to develop the algorithm for pattern search in any number of dimensions in the form of algebraic equations and inequalities suitable for numerical computation. The application of these expressions will then be illustrated by an example. As described informally above, the search proceeds in stages. Each stage is a pattern consisting of a set of successful pattern moves terminated by an unsuccessful move. After each stage the search is restarted in a new pattern from the final point reached in the previous pattern. The algorithm for the search will be concerned with the movements made in a single pattern. The overall search consists of a series of such patterns.

The search starts from a first approximation, $\mathbf{x}_0$, to the solution. Also necessary are a set of increments to be used in local explorations around the first approximation. It will be simplest to denote the increments of $\mathbf{x}$ by $\mathbf{y} = [y_1 \; y_2 \; . . . y_n]^T$. In order to terminate the search a lower limit to the increment size is necessary. Since increments are halved to improve resolution, an integer number K can be specified so that when the increments have been halved K times the increment size is reduced by a factor $(\frac{1}{2})^K$ and the minimum should be located within the resolution of the resulting increment length. Recall that the actual movement made in a pattern move is equal to the algebraic sum, in the direction of each axis, of movements made in each local exploration since the start of the pattern. In order to keep a record of the current multiple of the initial increments of movement, a column vector **N** whose elements correspond to axis directions is maintained. The current value of an element of **N** gives the number of initial increments of movement required in the corresponding direction for the next pattern move.

Two tests are also required. The first is a test for *undefined direction* made at the start of a new pattern when local exploration is carried out to determine the initial direction of movement. If all test positions give a greater value of E than the first approximation, the increment size must be halved in an attempt to improve resolution. After K such halvings the search is terminated.

The second test is made for *rejection of a new base* about to be established by a pattern move. If local exploration at this new point gives a least value E which is greater than that obtained in the previous local exploration then the projected move is a failure. The search then reverts back to the previous best point and a new pattern is commenced.

It will be instructive to describe the method in step by step form at this point, leaving the details of procedures required until afterwards. In the following description k counts the number of times the increments are halved, t is the number of pattern moves made to date in the current pattern (in effect the iteration number), and the vector $\mathbf{x}_A$ stores the best point obtained in the local exploration about the point $\mathbf{x}$.

Pattern search

Step: 1 Input data f, $\mathbf{x}_0$, $\mathbf{y}$, K, and set k = 0.

2 Begin a new pattern move and set $\mathbf{N} = 0$.
Test for end of optimization.
If k = K, terminate the search. The answer is then given by the current value of $\mathbf{x}$.

3. Make a local exploration about $\mathbf{x}_0$.
Calculate $\mathbf{N}$ which then defines a direction for movement.
Test for undefined direction.
If $\mathbf{N} = 0$, set $\mathbf{y} = \frac{1}{2}\mathbf{y}$, set $k = k + 1$, and go to Step 2.
Otherwise, the best point in the local exploration is given by $\mathbf{x}_{A,0} = \mathbf{x}_0 + \mathbf{N}\,\mathbf{y}$.

4. Make the first pattern move to the new base $\mathbf{x}_1$, where $\mathbf{x}_1 = \mathbf{x}_{A,0} + \mathbf{N}\mathbf{y}$.
Set $t = 1$.

5. Make a local exploration at base $\mathbf{x}_t$ to find the locally best point $\mathbf{x}_{A,t}$.
Update $\mathbf{N}$ by adding the number of increments used.

6. Test for rejection of base $\mathbf{x}_t$.
If $f(\mathbf{x}_{A,t}) \geqslant f(\mathbf{x}_{A,t-1})$, set $\mathbf{x}_0 = \mathbf{x}_{A,t} - \mathbf{N}\mathbf{y}$, and go to Step 2.

7. Test for undefined direction.
If $\mathbf{N} = 0$, set $\mathbf{y} = \frac{1}{2}\mathbf{y}$, set $k = k + 1$, and go to Step 2.

8. Make a pattern move to $\mathbf{x}_{t+1}$, where $\mathbf{x}_{t+1} = \mathbf{x}_{A,t} + \mathbf{N}\mathbf{y}$.
Set $t = t + 1$, and go to Step 5.

All necessary vector equations for making a move have been introduced in the step by step instructions. Each new value of $\mathbf{x}$ is a combination of the old value plus a number of increments defined by $\mathbf{N}$. It remains to give the details of local exploration and the updating of $\mathbf{N}$.

It is possible to use a single vector for the successive $\mathbf{x}$ positions derived from $\mathbf{x}_t$, $\mathbf{x}_{A,t}$, and $\mathbf{x}_{t+1}$. The local exploration is used to derive $\mathbf{x}_{A,t}$ from $\mathbf{x}_t$ and in the process the vector $\mathbf{N}$ is updated. The progressive change from $\mathbf{x}_t$ to $\mathbf{x}_{A,t}$ is carried out in each direction in turn. Denoting an intermediate value of x by $\mathbf{x}_{t,i-1}$ (beginning with $\mathbf{x}_{t,0} = \mathbf{x}_t$) the following search procedure is used in the ith direction.

Evaluate the function at $\mathbf{x}_{t,i-1}$ and at a point removed from $\mathbf{x}_{t,i-1}$ by an amount y_i in the direction of the ith axis. If $f(\mathbf{x}_{t,i-1})$ is greater than $f(\mathbf{x}_{t,i-1} + y_i \mathbf{1}_i)$, where $\mathbf{1}_i$ is the ith unit vector, label the new point $\mathbf{x}_{t,i}$ and increase the ith element of **N** by 1. In the contrary case, evaluate f at a position removed from $\mathbf{x}_{t,i-1}$ by an amount $-y_i$ in the direction of the ith axis. If $f(\mathbf{x}_{t,i-1})$ is greater than $f(\mathbf{x}_{t,i-1} - y_i \mathbf{e}_i)$, label the new point $\mathbf{x}_{t,i}$ and decrease the ith element of **N** by 1. In the contrary case, set $\mathbf{x}_{t,i}$ equal to $\mathbf{x}_{t,i-1}$ and leave **N** unchanged. This process is repeated for all i from 1 to n. **N** is progressively updated as $\mathbf{x}_t$ is progressively changed to $\mathbf{x}_{A,t}$. The pattern move to $\mathbf{x}_{t+1}$ can then be carried out according to the formula in Step 8 without further modifications to the vectors needed.

Example 4. Minimize the function given by $f(x_1, x_2) = 8x_1^2 - 4x_1x_2 + 5x_2^2$ by pattern search with $\mathbf{x}_0 = [5\ 2]^T$, $\mathbf{y} = [0.6\ 0.3]^T$, and $K = 2$.

This quadratic function has elliptical contours and is therefore simple to minimize. It is obvious that the solution is at the point (0,0). Table 4 showing the organization of the computation was compiled using a slide rule. The effects of the tests in the step by step procedure are illustrated in the table.

At Iteration 4 the increase in size of the pattern move has built up and the move has overstepped the central area of the valley. The search therefore reverts to the best point found during Iteration 3. At Iteration 5 the local exploration fails to establish a new direction and $\mathbf{N} = [0\ 0]^T$. Step size is therefore halved. The direction of movement established at Iterations 6 and 7 is rejected by the local exploration at Iteration 8. The search returns to the minimum point obtained in Iteration 7 with $\mathbf{N} = [0\ 0]^T$. Resolution is again increased by reducing step size. Since this is then equal to the specified minimum step size, the search is terminated and the point given by $\mathbf{x} = [-0.1\ -0.1]^T$ designated the minimum. Since the true minimum is at the point (0,0), further reductions in step size would be required to approach it more closely.

Table 4. Pattern search

Iteration t	0	1	2	3	4	5	6	7	8
Initial values									
$\mathbf{x}_t$	$\begin{bmatrix}5.0\\2.0\end{bmatrix}$	$\begin{bmatrix}3.8\\1.4\end{bmatrix}$	$\begin{bmatrix}2.0\\1.1\end{bmatrix}$	$\begin{bmatrix}-0.4\\0.2\end{bmatrix}$	$\begin{bmatrix}-1.0\\-0.4\end{bmatrix}$	$\begin{bmatrix}0.2\\0.2\end{bmatrix}$	$\begin{bmatrix}0.2\\0.2\end{bmatrix}$	$\begin{bmatrix}0.2\\-0.1\end{bmatrix}$	$\begin{bmatrix}-0.4\\-0.25\end{bmatrix}$
$\mathbf{y}$	$\begin{bmatrix}0.6\\0.3\end{bmatrix}$	$\begin{bmatrix}0.6\\0.3\end{bmatrix}$	$\begin{bmatrix}0.6\\0.3\end{bmatrix}$	$\begin{bmatrix}0.6\\0.3\end{bmatrix}$	$\begin{bmatrix}0.6\\0.3\end{bmatrix}$	$\begin{bmatrix}0.6\\0.3\end{bmatrix}$	$\begin{bmatrix}0.3\\0.15\end{bmatrix}$	$\begin{bmatrix}0.3\\0.15\end{bmatrix}$	$\begin{bmatrix}0.3\\0.15\end{bmatrix}$
k	0	0	0	0	0	0	1	1	1
Exploration in x_1 direction									
$f(\mathbf{x}_t)$	180	102	37	1.8	7.2	0.36	0.36	0.45	1.19
$f(\mathbf{x}_t + y_1 \mathbf{e}_1)$	226	140	49	0.36	1.44	4.68	1.8	2.25	0.25
$f(\mathbf{x}_t - y_1 \mathbf{e}_1)$	140	74	15.6			1.8	0.36	0.09	
Intermediate values									
$\mathbf{N}$	$\begin{bmatrix}-1\\0\end{bmatrix}$	$\begin{bmatrix}-2\\-1\end{bmatrix}$	$\begin{bmatrix}-3\\-1\end{bmatrix}$	$\begin{bmatrix}-2\\-2\end{bmatrix}$	$\begin{bmatrix}-1\\-2\end{bmatrix}$	$\begin{bmatrix}0\\0\end{bmatrix}$	$\begin{bmatrix}0\\0\end{bmatrix}$	$\begin{bmatrix}-1\\-1\end{bmatrix}$	$\begin{bmatrix}0\\-1\end{bmatrix}$
$\mathbf{x}_{t,1}$	$\begin{bmatrix}4.4\\2.0\end{bmatrix}$	$\begin{bmatrix}3.2\\1.4\end{bmatrix}$	$\begin{bmatrix}1.4\\1.1\end{bmatrix}$	$\begin{bmatrix}0.2\\0.2\end{bmatrix}$	$\begin{bmatrix}-0.4\\-0.4\end{bmatrix}$	$\begin{bmatrix}0.2\\0.2\end{bmatrix}$	$\begin{bmatrix}0.2\\0.2\end{bmatrix}$	$\begin{bmatrix}-0.1\\-0.1\end{bmatrix}$	$\begin{bmatrix}-0.1\\-0.25\end{bmatrix}$
Exploration in x_2 direction									
$f(\mathbf{x}_{t,1})$	140	73.8	15.6	0.36	1.44	0.36	0.36	0.09	0.25
$f(\mathbf{x}_{t,1} + y_2 \mathbf{e}_2)$	141	74.6	17.6	1.2	1.17	1.23	0.65	0.11	0.09
$f(\mathbf{x}_{t,1} - y_2 \mathbf{e}_2)$	139	73.9	14.4	0.45		0.45	0.29	0.29	
Final values									
$\mathbf{N}$	$\begin{bmatrix}-1\\-1\end{bmatrix}$	$\begin{bmatrix}-2\\-1\end{bmatrix}$	$\begin{bmatrix}-3\\-2\end{bmatrix}$	$\begin{bmatrix}-2\\-2\end{bmatrix}$		$\begin{bmatrix}0\\0\end{bmatrix}$	$\begin{bmatrix}0\\-1\end{bmatrix}$	$\begin{bmatrix}-1\\-1\end{bmatrix}$	$\begin{bmatrix}0\\0\end{bmatrix}$
$\mathbf{x}_{A,t}$	$\begin{bmatrix}4.4\\1.7\end{bmatrix}$	$\begin{bmatrix}3.2\\1.4\end{bmatrix}$	$\begin{bmatrix}1.4\\0.8\end{bmatrix}$	$\begin{bmatrix}0.2\\0.2\end{bmatrix}$			$\begin{bmatrix}0.2\\0.05\end{bmatrix}$	$\begin{bmatrix}-0.1\\-0.1\end{bmatrix}$	$\begin{bmatrix}-0.1\\-0.1\end{bmatrix}$
					(i)	(ii)			(iii)

Note (i) $f(\mathbf{x})$ too large in region of $\mathbf{x}_4$. Use $\mathbf{x}_{A,3}$ as base for a new pattern and set $\mathbf{N} = [0\ 0]^{\mathrm{T}}$.

(ii) Undefined direction. Halve increments and set $k = 1$.

(iii) Undefined direction. Halve increments and set $k = 2$.

Since $k = K$ terminate the search with $\mathbf{x} = [-0.1\ -0.1]^{\mathrm{T}}$

3.3 Gradient methods

Gradient methods for optimization are based on the Taylor expansion given by equation 1.17. The expansion is repeated here with the terms involving third and higher order derivatives neglected, i.e.

$$f(\mathbf{x} + \mathbf{\Delta x}) \approx f(\mathbf{x}) + \mathbf{g}^T \mathbf{\Delta x} + \tfrac{1}{2} \mathbf{\Delta x}^T \mathbf{H} \mathbf{\Delta x}. \qquad (3.1)$$

The last two terms on the right hand side are a scalar correction to the function value at $\mathbf{x}$ which yields an approximation to the function value at $\mathbf{x} + \mathbf{\Delta x}$. We shall denote such a correction in general by ΔE so that

$$f(\mathbf{x} + \mathbf{\Delta x}) \approx E + \Delta E. \qquad (3.2)$$

Optimization methods which neglect the final term in Equation 3.1, using only the Jacobian gradient vector $\mathbf{g}$ to calculate ΔE, are termed *first order* methods. Second derivatives are then assumed small enough to be neglected. If an optimization method utilizes second derivatives it is termed a *second order* method.

In general, experience has shown that gradient methods are superior to search methods if the functions involved have continuous derivatives which can be evaluated analytically. For functions with discontinuous derivatives such as those in Figure 3, however, search methods may be essential. It could be said that the art of optimization lies in the appropriate choice of method for the problem to be solved.

Steepest descent

The *steepest descent* method uses the Jacobian gradient g to determine a suitable direction of movement. It is the fundamental first order method.

The first approximation to the optimum defines a point at which the function is evaluated to yield E and a suitable change ΔE is found by evaluation of the Jacobian $\mathbf{g}$. The effect on E of a small change $\mathbf{\Delta x}$ in $\mathbf{x}$ is given to the first order of approximation by

$$\Delta E = \mathbf{g}^{\mathrm{T}}\ \mathbf{\Delta x} = \Sigma_{i=1}^{n} \frac{\partial f(\mathbf{x})}{\partial x_i} \Delta x_i\ , \tag{3.3}$$

where $\Delta x_1, \ldots, \Delta x_n$ are the elements of $\mathbf{\Delta x}$.
This equation can be thought of as involving the scalar, or dot, product of two vectors. The *scalar product* of two vectors **u** and **v** is defined as

$$\mathbf{u}\cdot\mathbf{v} = \Sigma_{i=1}^{n} u_i\, v_i = |\mathbf{u}||\mathbf{v}| \cos\theta, \tag{3.4}$$

where θ is the angle between the two vectors and the magnitude of a vector, e.g. **x**, is given by $|\mathbf{x}|$. The (*Euclidian*) *magnitude* or *norm* $|\mathbf{x}|$ of **x** is a number given by the square root of the sum of the squares of the elements of **x**, i.e.

$$|\mathbf{x}| = \sqrt{(\Sigma_{i=1}^{n} x_i^2)} = \sqrt{(\mathbf{x}^{\mathrm{T}}\mathbf{x})} = \sqrt{(\mathbf{x}\cdot\mathbf{x})}.$$

If follows from these definitions and 3.3 that ΔE is given by

$$\Delta E = |\mathbf{g}||\mathbf{\Delta x}| \cos\theta. \tag{3.5}$$

Clearly for fixed magnitudes $|\mathbf{g}|$ and $|\mathbf{\Delta x}|$, ΔE depends on $\cos\theta$, taking a maxiumum positive value if $\theta = 0$ and a maximum negative value if $\theta = \pi$. The maximum reduction in E therefore occurs if $\theta = \pi$, from which it follows that the minimising change $\mathbf{\Delta x}$ in **x** should be in the direction of the negative gradient $-\mathbf{g}$.

Now the unit vector in the direction of $-\mathbf{g}$ is given by **u** where

$$\mathbf{u} = \frac{-\mathbf{g}}{|\mathbf{g}|}, \tag{3.6}$$

so that the change $\mathbf{\Delta x}$ to be made in **x** is proportional to **u**, i.e.

$$\mathbf{\Delta x} = \lambda\mathbf{u}. \tag{3.7}$$

A single parameter or linear search of the function of λ given by $f(\mathbf{x} + \lambda\mathbf{u})$ determines the optimum value for λ. This linear search yields the restricted minimum of the function in the direction of **u**. Since the direction of the negative gradient is

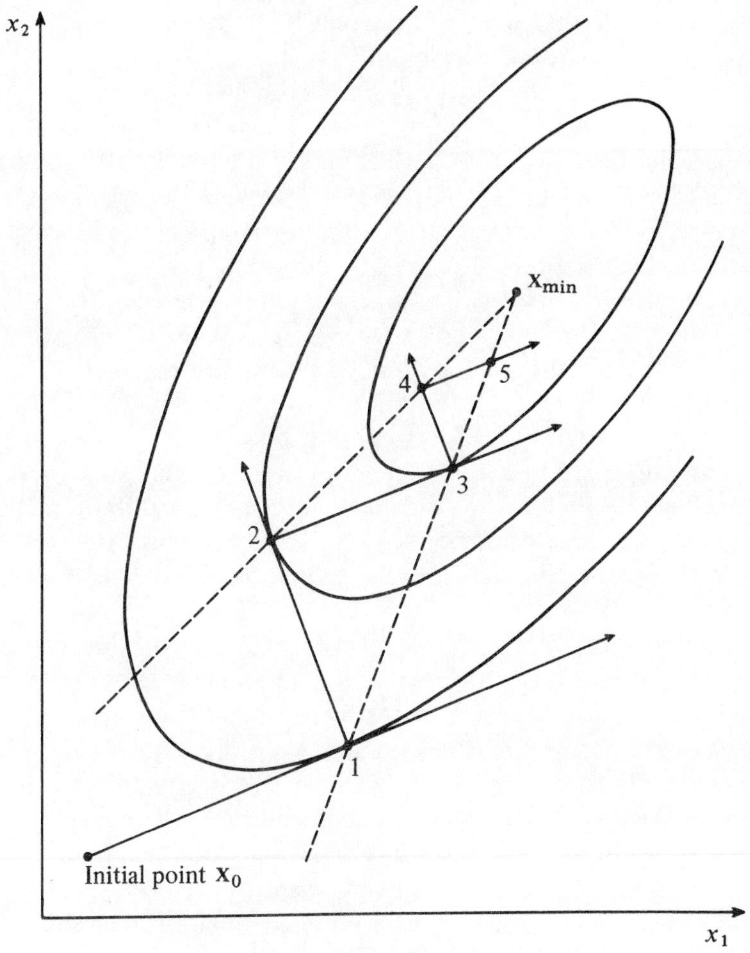

Fig. 15. *Steepest descent minimization.*

in general not toward the (unrestricted) minimum, the process must then be repeated iteratively. A typical minimization of a function of two variables with elliptical contours is depicted in Figure 15. The term 'steepest descent' arises since, as we have seen, descent along the negative gradient maximizes the reduction in E generated by first order approximation.

The steepest descent minimization of Figure 15 follows a zig-zag path rather similar to the one at a time search of Figure 10. Progress of the minimization near to optimum is very slow due to the very large number of steps required. Progress from a distant point toward the optimum is much better but depends on proper scaling for best results. It should be noticed that in the case of the elliptical contours shown in Figure 15 the zig-zag path lies between two straight lines which meet at the optimum. The method of *parallel tangents,* or the *partan* method, due to Shah, Buehler, and Kempthorne (1964) is developed from this observation (see also Wilde (1964), Chapter 5). Further, if λ is found exactly, successive path directions are at right angles since at the optimum λ,

$$\partial/\partial\lambda f(\mathbf{x} + \lambda\mathbf{u}) = \nabla f(\mathbf{x} + \lambda\mathbf{u})\mathbf{u} = 0.$$

Equations 3.3 to 3.7 for steepest descent minimization were developed in vector form and are suitable for use with functions of any number of parameters. Since the method incorporates a search in one dimension in place of the fixed incremental steps used by search methods, the minimization must be terminated differently. The test for terminating this linear search can take many different forms and must often be tailored to the problem. It normally involves the recognition that over a number of successive iterations the value E remains almost constant. The test suggested here is that the procedure should detect a slowing down of the minimization by the condition $|\lambda_k \mathbf{u}_k| \leqslant |K\lambda_0 \mathbf{u}_0|$, i.e. $\lambda_k \leqslant K\lambda_0$.

The factor K, which lies between 0 and 1, has been introduced so that the minimization is terminated when the step $\lambda_k \mathbf{u}_k$ taken at iteration k is less than the fraction K of the first step taken. Initial step sizes are of course not specified *a priori*. Note that λ_{k-1} is used instead of λ_k in practice, since K has been incremented by one at the point of test.

The steepest descent optimization can now be put into step by step form.

Steepest descent minimization

Step: 1. Input data f, $\mathbf{x}_0$, and K. Set $k = 0$

Step: 2. Evaluate $E_k = f(\mathbf{x}_k)$.
Test for the end of the minimization.
If $\lambda_{k-1} \leqslant K\,\lambda_0$, terminate the minimization and output $\mathbf{x}_{\min} = \mathbf{x}_k$, $E_{\min} = E_k$.
3. Evaluate the gradient $\mathbf{g}_k$ at point $\mathbf{x}_k$.
4. Compute the search direction $\mathbf{u}_k = -\mathbf{g}_k/|\mathbf{g}_k|$.
5. Perform a linear search, i.e. a single parameter minimization, in the search direction to find λ_k.
6. Generate a new point $\mathbf{x}_{k+1} = \mathbf{x}_k + \lambda_k \mathbf{u}_k$.
Set $k = k + 1$ and go to Step 2.

Before working an example and passing on to the study of second order methods it is worthwhile to investigate possible improvements to the steepest descent method should the Hessian matrix $\mathbf{H}$ be available.

If expression 3.7, giving $\mathbf{\Delta x}$ as $-\lambda\mathbf{g}/|\mathbf{g}|$, is substituted into equation 3.1 then

$$\Delta E = -\frac{\mathbf{g}^{\mathrm{T}}\lambda\mathbf{g}}{|\mathbf{g}|} + \frac{\lambda^2}{2}\frac{\mathbf{g}^{\mathrm{T}}\mathbf{H}\mathbf{g}}{|\mathbf{g}|^2} \tag{3.8}$$

i.e.

$$\Delta E = -\lambda|\mathbf{g}| + \frac{\lambda^2}{2}\frac{\mathbf{g}^{\mathrm{T}}\mathbf{H}\mathbf{g}}{|\mathbf{g}|^2}. \tag{3.9}$$

At the minimum of the function in the direction $\mathbf{g}/|\mathbf{g}|$, variation of λ must give a maximum negative value of ΔE. Therefore, differentiating with respect to λ and equating the result to zero,

$$\frac{\partial(\Delta E)}{\partial\lambda} = -|\mathbf{g}| + \lambda\frac{\mathbf{g}^{\mathrm{T}}\mathbf{H}\mathbf{g}}{|\mathbf{g}|^2} = 0, \tag{3.10}$$

i.e.

$$\lambda = \frac{|\mathbf{g}|^3}{\mathbf{g}^{\mathrm{T}}\mathbf{H}\mathbf{g}}. \tag{3.11}$$

The value of λ can therefore be determined without a linear search when $\mathbf{H}$ is known.

It is possible to develop these considerations even further. If a guess μ is made to the value of λ, then from equation 3.1

$$f\left(\mathbf{x} - \frac{\mu\mathbf{g}}{|\mathbf{g}|}\right) = f(\mathbf{x}) - \frac{\mathbf{g}^T \mu\mathbf{g}}{|\mathbf{g}|} + \frac{\mu^2}{2} \frac{\mathbf{g}^T \mathbf{H}\mathbf{g}}{|\mathbf{g}|^2}. \quad (3.12)$$

Solving for $\mathbf{g}^T \mathbf{H}\mathbf{g}$ and substituting into equation 3.11 gives

$$\lambda, = \frac{\mu^2 |\mathbf{g}|}{2} \left\{ f(\mathbf{x} - \frac{\mu\mathbf{g}}{|\mathbf{g}|} - f(\mathbf{x}) + \mu|\mathbf{g}| \right\}^{-1}. \quad (3.13)$$

At the expense of making a guess to the value of λ and an extra function evaluation it has been possible to eliminate the Hessian altogether. At the kth iteration the obvious choice for μ is λ_{k-1}. The steepest descent method in this case remains a first order method without the necessity of a linear search. Rohrer (see Temes (1969)) reports good results with this modification.

Example 5. Minimize the function given by $f(x_1, x_2) = 8x_1^2 - 4x_1x_2 + 5x_2^2$ using the method of steepest descent. Use an initial approximation $\mathbf{x}_0 = [5\ 2]^T$. Terminate the search using $K = 0.1$.

This example concerns a quadratic function which is clearly continuous and differentiable. Since the graph of this function is a paraboloid in three dimensional space, planar sections of it are either elliptical or parabolic. It follows that the contours of the function, i.e. horizontal sections, are elliptical, while the curve described by evaluating the function along a line is a parabola. The latter is easily verified algebraically by evaluating f at $\mathbf{x} + \lambda\mathbf{u}$. Linear search for λ, the distance to be moved in the direction of the negative gradient, will therefore be carried out using a quadratic curve fitting technique.

E is given exactly by a quadratic function of λ, to be determined, i.e.

$$E = a\lambda^2 + b\lambda + c.$$

If the three values of λ are chosen and f is evaluated at the corresponding values of x_1 and x_2, the constants a, b, and c can be determined. Values for λ of 0, 1, and 2 yield

$$E_0 = c \qquad \text{for } \lambda = 0$$

$$E_1 = a + b + c \qquad \text{for } \lambda = 1$$

$$E_2 = 4a + 2b + c \qquad \text{for } \lambda = 2,$$

which can be solved for a, b, and c. The values of λ for the minimum value of E is derived as usual by differentiation,

$$\frac{\partial E}{\partial \lambda} = 2a\lambda + b = 0,$$

i.e.

$$\lambda = \frac{-b}{2a}.$$

Note also that the elements of the Jacobian g are given by

$$\frac{\partial f(\mathbf{x})}{\partial x_1} = 16x_1 - 4x_2, \quad \frac{\partial f(\mathbf{x})}{\partial x_2} = -4x_1 + 10x_2.$$

The minimization for this example can now be organised into tabular form in which progress is easily traced.

Table 5. Steepest descent minimization

Iteration k	0	1	2
$\mathbf{x}_k^T$	[5 2]	[0.5 2]	[0.5 0.2]
$\mathbf{g}_k^T$	[72 0]	[0 18]	[7.2 0]
$\mathbf{u}_k^T$	[−1 0]	[0 −1]	[−1 0]
E_0	180	18	1.8
E_1	116	5	2.6
E_2	68	2	19.4
a	8	5	8
b	−72	−18	−7.2
c	180	−18	1.8
λ_k	4.5	1.8	0.45
$\mathbf{x}_{k+1}^T$	[0.5 2]	[0.5 0.2]	[0.05 0.2]

Search terminated since $K\lambda_0 = \lambda_2 = 0.45$.
$\mathbf{x}_{\min}^T = [0.05 \;\; 0.2]$, $E_{\min} = 0.18$.

In this example, which for purposes of direct comparison is the same one used to illustrate pattern search, it is fortuitous that the first approximation to $\mathbf{x}_{\min}$ at the point $[5\ 2]^T$ leads to gradient vectors parallel to the co-ordinate axes. Therefore the example is easy to follow numerically. The successive search directions $\mathbf{u}$, for example, are clearly at right angles, another indication that λ has been found exactly. Since the linear searches for λ are carried out parallel to the axes, this example also illustrates the one at a time search. In one at a time search, the search directions are specifically parallel to the axes and not derived from the gradient. For the initial value of $\mathbf{x}$ used in this example the similarity between the two methods is over-emphasised. It is instructive to repeat the example for both methods from various first approximations to $\mathbf{x}_{\min}$.

Second order methods

Improved gradient techniques rely on the provision of the Hessian matrix $\mathbf{H}$ of second order partial derivatives. Some of the advanced methods discussed in Chapter 4 make approximations to the Hessian instead of using direct computation. Once the approximation is made however the methods reduce to the fundamental techniques of this chapter.

The Taylor expansion of equation 3.1 can be used to approximate the minimum value of the objective function from points $\mathbf{x}$ near to the minimum $\mathbf{x}_{\min}$ as

$$f(\mathbf{x}_{\min}) \approx f(\mathbf{x}) + \mathbf{g}^T \mathbf{\Delta x} + \tfrac{1}{2}\, \mathbf{\Delta x}^T \mathbf{H} \mathbf{\Delta x}, \tag{3.14}$$

where $\mathbf{x}_{\min} = \mathbf{x} + \mathbf{\Delta x}$ and the Jacobian gradient vector $\mathbf{g}$ and the Hessian matrix $\mathbf{H}$ are evaluated at the point $\mathbf{x}$. Our object is to determine the elements $\mathbf{\Delta} x_i\ (i = 1, \ldots, n)$ of the movement $\mathbf{\Delta x}$ required to approximate the minimum from the point $\mathbf{x}$.

Putting expression 3.14 into co-ordinate form yields

$$f(\mathbf{x}_{\min}) \approx f(\mathbf{x}) + \sum_{i=1}^{n} \frac{\partial f(\mathbf{x})}{\partial x_i}\, \Delta x_i + \frac{1}{2} \sum_{i=1}^{n} \sum_{j=1}^{n} \Delta x_i \frac{\partial^2 f(\mathbf{x})}{\partial x_i\, \partial x_j}\, \Delta x_j \tag{3.15}$$

To determine $\mathbf{\Delta x}$ approximately, consider $\mathbf{g}$ and $\mathbf{H}$ fixed and partially differentiate 3.15 with respect to the elements Δx_j for each j from 1 to n. Setting the results to zero gives

$$\frac{\partial f(\mathbf{x})}{\partial x_j} + \sum_{i=1}^{n} \Delta x_i \frac{\partial^2 f(\mathbf{x})}{\partial x_i \, \partial x_j} = 0 \quad j = 1, \ldots, n \quad (3.16)$$

for the first order condition for a minimum of the right hand side of 3.15 in terms of the x_j. In matrix form 3.16 is

$$\mathbf{g} = -\mathbf{H}\ \mathbf{\Delta x} \quad (3.17)$$

Solving 3.17 yields

$$\mathbf{\Delta x} = -\mathbf{H}^{-1}\mathbf{g} \quad (3.18)$$

as the approximation to the required movement to the minimum $\mathbf{x}_{\mathbf{min}}$ from a point $\mathbf{x}$ near to the minimum (the current point). Since second order derivatives are constant for quadratic functions, the Hessian matrix $\mathbf{H}$ at any point $\mathbf{x}$ is equal to the Hessian matrix $\mathbf{H}_{\mathbf{min}}$ evaluated at the minimum. For quadratic functions the approximation is therefore exact and the minimum can be reached from any $\mathbf{x}$ in a single step. In general the required movement to a minimum from a nearby point is given approximately by the equation 3.18.

Equation 3.18 is fundamental to all second order solutions to the minimization problem. When it is used directly to generate successive movements toward a minimum from a given inital value $\mathbf{x}_0$ the method is known as the generalized *Newton-Raphson* method (or simply as *Newton's* method). Direct use of equation 3.18 is limited however because the Hessian matrix $\mathbf{H}$ must be computed and inverted at each step of any iterative procedure. Furthermore, considering the Taylor series expansion of f about $\mathbf{x}_{\mathbf{min}}$ (analogous to 3.14), viz.

$$f(\mathbf{x}) = f(\mathbf{x}_{\mathbf{min}}) - \mathbf{g}_{\mathbf{min}}^{\mathrm{T}}\ \mathbf{\Delta x} + \tfrac{1}{2}\,\mathbf{\Delta x}^{\mathrm{T}}\ \mathbf{H}_{\mathbf{min}}\ \Delta\mathbf{x},$$

and using the fact that the Jacobian gradient vector of a function with continuous partial derivatives is zero at the minimum point, i.e. $\mathbf{g}_{\mathbf{min}} = 0$, we have that for any $\Delta\mathbf{x} \neq 0$,

$$\Delta \mathbf{x}^T \mathbf{H}_{min} \Delta \mathbf{x} = 2[f(\mathbf{x}) - f(\mathbf{x}_{min})] > 0, \quad (3.19)$$

i.e. $\mathbf{H}_{min}$ is *positive definite*. However, $\mathbf{H}$ is not necessarily positive definite, since it is evaluated at a point other than the minimum, so that the process may not converge. This situation is most likely to occur at some distance from the minimum, since at a point close to the minimum all sufficiently differentiable functions tend to become quadratic as third order terms in their Taylor expansions about the minimum become negligible. In practice, second order methods are sometimes preceded by first order methods, often steepest descent, away from the minimum. As the minimum is approached a more powerful second order technique is used. Instead of computing $\mathbf{H}^{-1}$ however, a positive definite approximation to it is usually maintained so that convergence can be guaranteed. Advanced second order gradient methods of this type are discussed further in Chapter 4.

Least squares

The application of optimization techniques to curve fitting by *least squares* is best considered as a method for minimizing the error between sample points on the graphs of two functions of an independent variable t. One of these functions depends on several parameters, taken to be elements of a vector $\mathbf{x}$, which are to be chosen. Such a situation arises, for example, through regression problems in statistics and econometrics. The *errors of fit* or *residuals* at each of a series $t_1, t_2, \ldots, t_m$ of m data points (usually the difference between the function values at each point) may be expressed as a series of evaluations of a function f at t_k, $f(\mathbf{x}, t_1), f(\mathbf{x}, t_2), \ldots, f(\mathbf{x}, t_m)$, which are dependent on the parameters $\mathbf{x}$ of the approximating function. Of course m must at least equal the number of parameters, n, in $\mathbf{x}$. As discussed in Section 1.2, total error is then a function of the errors at the individual data points, i.e.

$$E = g\{f(\mathbf{x},t_1), f(\mathbf{x},t_2), \ldots, f(\mathbf{x},t_m)\}. \quad (3.20)$$

If the functions $f(\mathbf{x}, t_k)$ are rewritten $f_k(\mathbf{x})$ and the function g is replaced by a sum of squares, we have

$$E = \Sigma_{k=1}^{m} [f_k(\mathrm{x})]^2. \tag{3.21}$$

Each function $f_k(\mathrm{x})$ represents a component of the total error E and if a vector valued function **F** is defined by

$$F^{\mathrm{T}} = [f_1(\mathrm{x}) f_2(\mathrm{x}) \ldots f_m(\mathrm{x})], \tag{3.22}$$

$$E = \mathbf{f}^{\mathrm{T}}\mathbf{f}. \tag{3.23}$$

The contribution to the value of E due to changes in the parameters $x_1, x_2, \ldots, x_n$ defines the Jacobian gradient vector. It is clear that the change in one parameter x_i will affect all the elements of **F** in equation 3.22 and that each of these will contribute to the total error in equation 3.23. It is therefore convenient to define a new matrix **A** which gives the variation of each function $f_1(\mathrm{x}), \ldots, f_m(\mathrm{x})$ due to variation of each parameter $x_1, \ldots, x_n$, viz.

$$\mathbf{A} = \begin{bmatrix} \dfrac{\partial f_1}{\partial x_1} & \dfrac{\partial f_1}{\partial x_2} & \cdots\cdots & \dfrac{\partial f_1}{\partial x_n} \\ \dfrac{\partial f_2}{\partial x_1} & \dfrac{\partial f_2}{\partial x_2} & \cdots\cdots & \dfrac{\partial f_2}{\partial x_n} \\ \vdots & & & \vdots \\ \dfrac{\partial f_m}{\partial x_1} & \cdots\cdots & \cdots\cdots & \dfrac{\partial f_m}{\partial x_n} \end{bmatrix} \tag{3.24}$$

Since m usually exceeds n this matrix is not in general square.

The elements of the Jacobian **g** can be derived by partial differentiation of equation 3.21 with respect to each parameter in turn as

$$\frac{\partial E}{\partial x_i} = \Sigma_{k=1}^{m} 2[f_k(\mathrm{x})] \frac{\partial f_k(\mathrm{x})}{\partial x_i}. \tag{3.25}$$

Hence

$$\begin{bmatrix} \frac{\partial E}{\partial x_1} \\ \vdots \\ \frac{\partial E}{\partial x_n} \end{bmatrix} = 2 \begin{bmatrix} \frac{\partial f_1}{\partial x_1} & \cdots\cdots & \frac{\partial f_m}{\partial x_1} \\ \vdots & & \vdots \\ \frac{\partial f_1}{\partial x_n} & \cdots\cdots & \frac{\partial f_m}{\partial x_n} \end{bmatrix} \begin{bmatrix} f_1(\mathbf{x}) \\ \vdots \\ f_m(\mathbf{x}) \end{bmatrix} \tag{3.26}$$

or in matrix form,

$$\mathbf{g} = 2\,\mathbf{A}^{\mathrm{T}}\,\mathbf{f}. \tag{3.27}$$

A second partial differentiation of equation 3.22, assuming that the f_k have continuous second partial derivatives (so that the order of partial differentiation can be interchanged), gives

$$\frac{\partial^2 E}{\partial x_i \partial x_j} = 2\,\Sigma_{k=1}^{m} \frac{\partial [f_k(\mathbf{x})]}{\partial x_i} \frac{\partial [f_k(\mathbf{x})]}{\partial x_j} + 2\,\Sigma_{k=1}^{m} f_k(\mathbf{x}) \frac{\partial^2 [f_k(\mathbf{x})]}{\partial x_i \partial x_j}. \tag{3.28}$$

The usual least squares procedure assumes that the second term can be neglected. Therefore,

$$\frac{\partial^2 E}{\partial x_i \partial x_j} \approx 2\,\Sigma_{k=1}^{m} \frac{\partial [f_k(\mathbf{x})]}{\partial x_i} \frac{\partial [f_k(\mathbf{x})]}{\partial x_j}. \tag{3.29}$$

This equation for all i and $j = 1, \ldots, n$ gives an approximation to the elements of the Hessian matrix as

$$\mathbf{H} \approx 2\,\mathbf{A}^{\mathrm{T}}\,\mathbf{A}. \tag{3.30}$$

The Hessian matrix and Jacobian vector can now be substituted into equation 3.18 to give the increment

$$\mathbf{\Delta x} = -\,[\mathbf{A}^{\mathrm{T}}\,\mathbf{A}]^{-1}\,\mathbf{A}^{\mathrm{T}}\,\mathbf{f}. \tag{3.31}$$

The resulting process is often termed the *Gauss-Newton* method.

In general $\mathbf{A}^{\mathrm{T}}\,\mathbf{A}$ is positive definite, so that this application of the Newton-Raphson method should converge. In some situations however it is important to limit the step size so that a solution is not predicted far outside the range of a valid first order approximation to **H**. In this case a fraction λ of the predicted change $\mathbf{\Delta x}$ is used and the process becomes

$$\mathbf{x}_{k+1} = \mathbf{x}_k + \lambda\,\mathbf{\Delta x}_k, \qquad (3.32)$$

where $\lambda < 1$ can either be fixed in advance or found by a linear search. This method is known as *damped least squares*.

In summary, equation 3.31 gives the least squares increment for the minimization of the sum of the squares of the m functions, $f_1(\mathbf{x}), \ldots, f_m(\mathbf{x})$ of n parameters. The least squares optimization procedure is the Newton-Raphson method utilizing this increment. Termination is effected by recognizing that the value of E remains almost constant over a number of successive iterations. The least squares minimization can now be put into step by step form.

Least squares

Step: 1. Input data $\mathbf{x}_0$, and the functions **f**. Set $k = 0$.

2. Evaluate $\mathbf{f}_k$ and E_k.
   If E_k has not reduced over a number of iterations, terminate the minimization and output $\mathbf{x}_{\mathbf{min}} = \mathbf{x}_k$ and $E_{\mathbf{min}} = E_k$.
3. Evaluate the $n \times m$ matrix **A**.
4. Derive the Jacobian $\mathbf{g} = 2\mathbf{A}^T\mathbf{f}$, and the Hessian $\mathbf{H} = 2\mathbf{A}^T\mathbf{A}$.
5. Compute $\mathbf{H}^{-1}$ and solve for $\mathbf{\Delta x} = -\mathbf{H}^{-1}\mathbf{g}$.
6. Generate a new point $\mathbf{x}_{k+1} = \mathbf{x}_k + \mathbf{\Delta x}$.
   Set $k = k + 1$, and return to Step 2.

Although the least squares method minimizes a special form of objective function, the following example also illustrates (rather trivially) the Newton-Raphson method.

Example 6. Minimize the value of $f(\mathbf{x}, t_1, t_2) = x_1 t_1^2 + x_2 t_2^2 - 1$ for values of the independent parameters (t_1, t_2) given by $(0, 1)$, $(\frac{1}{2}, \frac{1}{2})$, and $(1, 0)$ using the method of least squares.

Use an initial approximation $x_1 = 5$, $x_2 = 2$. Terminate the procedure when error reduction ceases.

The function to be minimized (after reordering terms for convenience) is given by

$$E = (x_2 - 1)^2 + (x_1 - 1)^2 + \left(\frac{x_1}{4} + \frac{x_2}{4} - 1\right)^2$$

In this form the minimization could be carried out by any of the search or gradient methods previously described, but we are here to consider the individual squared terms of E separately. The vector function $\mathbf{f}$ is therefore given by

$$\mathbf{f} = \begin{bmatrix} x_2 - 1 \\ x_1 - 1 \\ \dfrac{x_1}{4} - \dfrac{x_2}{4} - 1 \end{bmatrix}$$

Since $\mathbf{f}$ is linear in x_1 and x_2, the $\mathbf{A}$ matrix and consequently the Hessian $\mathbf{H}$ are constant. These can therefore be derived in advance of the main solution as

$$\mathbf{A} = \begin{bmatrix} 0 & 1 \\ 1 & 0 \\ \dfrac{1}{4} & \dfrac{1}{4} \end{bmatrix},$$

$$\mathbf{H} = 2\mathbf{A}^{\mathrm{T}}\mathbf{A} = 2\begin{bmatrix} \dfrac{17}{16} & \dfrac{1}{16} \\ \dfrac{1}{16} & \dfrac{17}{16} \end{bmatrix},$$

$$\mathbf{H}^{-1} = \frac{1}{2}\begin{bmatrix} \frac{17}{18} & \frac{-1}{18} \\ \frac{-1}{18} & \frac{17}{18} \end{bmatrix}.$$

The values for **A** and $\mathbf{H}^{-1}$ can be substituted into equation 3.31 to give an equation for $\Delta\mathbf{x}$ dependent only on **f**, i.e.

$$\Delta\mathbf{x} = -\mathbf{H}^{-1}\, 2\mathbf{A}^{\mathrm{T}}\, \mathbf{f}$$

$$= \begin{bmatrix} \frac{1}{18} & \frac{-17}{18} & \frac{-4}{18} \\ \frac{-17}{18} & \frac{1}{18} & \frac{-4}{18} \end{bmatrix} \mathbf{f}$$

Iterations of the least squares procedure can now be put into convenient tabular form.

Table 6. Least squares minimization

Iteration k	$\mathbf{x}_k$	$\mathbf{F}$	E	$\Delta\mathbf{x}$	$\mathbf{x}_{k+1}$
0	$\begin{bmatrix} 5 \\ 2 \end{bmatrix}$	$\begin{bmatrix} 1 \\ 4 \\ \frac{3}{4} \end{bmatrix}$	17.6	$\begin{bmatrix} -70/18 \\ -16/18 \end{bmatrix}$	$\begin{bmatrix} 20/18 \\ 20/18 \end{bmatrix}$
1	$\begin{bmatrix} 20/18 \\ 20/18 \end{bmatrix}$	$\begin{bmatrix} 2/18 \\ 2/18 \\ -8/18 \end{bmatrix}$	2.22	$\begin{bmatrix} 0 \\ 0 \end{bmatrix}$	$\begin{bmatrix} 20/18 \\ 20/18 \end{bmatrix}$

No further reduction in E, terminate the minimization.

$$\mathbf{x}_{\min} = [1.11 \quad 1.11]^{\mathrm{T}}, \; E_{\min} = 2.22$$

This example has been solved in one iteration only, the second iteration shows ideal behaviour by predicting zero values for $\Delta \mathbf{x}$ and the process is terminated. The reason for an exact solution, of course, is that the function to be minimized is quadratic and for quadratic functions the Newton-Raphson method always achieves the exact minimum in one step from any first approximation. With regard to the least squares method, the Hessian matrix derived directly by partial differentiation of the expression for E is equal to the approximation $2\mathbf{A}^T\mathbf{A}$ since total error is quadratic and its higher derivatives are zero. Such ideal behaviour should not be expected either in the exercises or in practice.

Exercises 3

1 Repeat example 5 using both steepest descent and one at a time search from a first approximation $\mathbf{x}_0 = [5 \quad 7]^T$.

2 Fit a section of the parabola $y^2 - a(x-b) = 0$ to four convenient points given by $y = \log_{10} x$ for $1 \leqslant x \leqslant 10$. Use the method of least squares with a first approximation given by $a = 0.1$ and $b = 1$.

3 Minimize the function given by

$$E = (x_2 - 1)^2 + (x_1 - 1)^2 + \left(\frac{x_1}{4} + \frac{x_2}{4} - 1\right)^2$$

from example 6 using the pattern search and steepest descent methods.

4 Minimize the function given by

$$E = [10 - (x_1 - 1)^2 - (x_2 - 1)^2]^2 + (x_1 - 2)^2 + (x_2 - 4)^2$$

from a first approximation $\mathbf{x}_0 = [1 \quad 0]^T$ by the steepest descent and pattern search methods. Plot progress on the contour graph prepared in answer to Exercise 2, Chapter 1. Also treat the squared terms in the function separately and minimize the sums of squares by the least squares method.

5 If access to a computer system is available, write subroutines for the pattern search, steepest descent, and least squares methods. Repeat the examples and exercises using the computer.

CHAPTER FOUR

Advanced methods

4.1 Introduction

The basic optimization methods of the previous chapters have been described in considerable detail since more advanced methods are developments of these techniques. The methods described in this chapter are in an introductory form suitable for further development by the reader, both numerically for use in practice, and mathematically, by reference to the research literature.

This chapter progresses through a description of seven main methods of unconstrained optimization in n dimensions to an introduction to curve fitting and function approximation with the minimax error criterion. The first mentioned are improved forms of the basic methods given in Chapter 3.

Numerical examples which illustrate the use of these advanced optimization methods are not given, since this is not the intended function of this chapter. At this stage the reader should be able to put the mathematical solutions into step by step form for numerical computation and to apply the result to the examples and exercises of Chapter 3. More difficult exercises at the end of this chapter illustrate some of the awkward function behaviour discussed in Chapter 1 and exhibited by practical problems. They begin to show the *necessity* of using a digital computer. It should be instructive to compare the methods described here with the methods used in existing library computer programs.

4.2 General considerations

A number of considerations have been advanced for evaluating the absolute and relative efficiencies of advanced optimization methods embodied in an operational computer code. These are both theoretical and practical. As is unfortunately the case in any rapidly developing field, there are always gaps in theoretical knowledge. More serious, there have in the past been false theoretical indications of efficiency due to certain properties of an algorithm which have been overlooked until computational experience accumulated. On the other hand there have also been false indications of general efficiency due to practical experience being limited to problems from a certain field of application. This is not to say, of course, that the best methods for problems in a special field may not be specially adapted algorithms. A specific instance of this proposition is the success of specially developed, *vis a vis* general, algorithms for the least squares problem. (In Section 4.4 we shall study two effective methods for this problem.) For the reasons given above it has become standard practice to test any new algorithm before publication on a battery of well known test problems (several of which are included in the exercises). It is important to include test problems with large n, as many methods which are tolerably efficient with few variables become totally unacceptable with many.

An efficient algorithm must be as simple as possible. Housekeeping operations – such as evaluating the function and its derivatives and solving linear equations at each iteration – and storage requirements in the computer should be minimized. However, in the design of an algorithm a balance must be struck between considerations of simplicity and the resulting speed of convergence.

An obvious preliminary indication of efficiency for an optimization method is a proof of theoretical convergence to a local minimum for a suitably broad class of functions. The force of the adjective 'theoretical' is that problems of numerical accuracy caused by the replacement of exact by machine arithmetic in the computer are ignored. The terms defined in the remainder of this paragraph are theoretical in the same sense. More important, but more difficult to establish than simple convergence, is a theoretical indication of the rate of convergence of an algorithm. For a function with continuous derivatives the method of steepest descent converges to a stationary point, see Curry (1944). As we saw in Section 3.3 the strategy of steepest descent is based on a first order Taylor series approximation to the function to be minimized. It is very slow near the minimum due to successive directions being at right angles. A technique which converges at the rate of the steepest descent method is said to exhibit *linear* convergence. An equivalent property for a convex function (see Figure 13), is that the ratio $|\mathbf{x}_{k+1} - \mathbf{x}_{\min}|/|\mathbf{x}_k - \mathbf{x}_{\min}|$ of the distance between the minimum and the current $\mathbf{x}$ at successive iterations converges to a positive constant less than one called the *convergence ratio.* For the method of steepest descent the convergence ratio can be shown to depend on the Hessian matrix at the minimum, $\mathbf{H}_{\min}$. We have noted that when a general function f has continuous second partial derivatives, $\mathbf{H}_{\min}$ must be a non-negative definite (symmetric) matrix at a local minimum. It is a result of linear algebra (see for example Birkhoff and Maclane, 1953) that its n *eigenvalues* μ_i, i.e. the solutions of the equations $\mathbf{H}_{\min}\,\mathbf{u}_i = \mu_i\mathbf{u}_i$ for suitable *eigenvectors* $\mathbf{u}_i$, $i = 1, \ldots, n$, are non-negative real numbers. For an f such that the maximum M and minimum m of these numbers are positive, it can be shown that the convergence ratio is given by

$$\beta = \left(\frac{M - m}{M + m}\right)^2 = \left(\frac{r - 1}{r + 1}\right)^2,$$

see for example Luenberger (1972), Chapter 7. Here $r = M/m$, the ratio of the largest to the smallest eigenvalue of $\mathbf{H}_{\min}$, is called the *condition number* of the matrix. Linear convergence is sometimes termed *geometric* convergence, since it follows from the definitions that for k' large enough

$$|\mathbf{x}_k - \mathbf{x}_{\min}| \approx \beta^{k-k'} |\mathbf{x}_{k'} - \mathbf{x}_{\min}| \, .$$

Note that as r tends to 1 from above, β tends to 0. Hence the more *ill-conditioned* $\mathbf{H}_{\min}$, i.e. the more r exceeds 1, the slower the rate of convergence of steepest descent. This is the theoretical foundation underlying the practical scaling of variables so that the derivatives of f are of approximately equal magnitudes. An algorithm with the property that $|\mathbf{x}_{k+1} - \mathbf{x}_{\min}|/|\mathbf{x}_k - \mathbf{x}_{\min}|$ converges to zero is said to exhibit *super-linear* convergence. If further the ratio $|\mathbf{x}_{k+1} - \mathbf{x}_{\min}|/|\mathbf{x}_k - \mathbf{x}_{\min}|^2$ converges to a constant, it exhibits *quadratic* or *second order* convergence. Since it is based upon the second order Taylor series expansion of the function about its minimum, when f possesses third derivatives the basic Newton-Raphson method may be shown to possess this property close to the minimum, see Luenberger, *op. cit.* In general, p^{th} *order* convergence can be defined for $p \geqslant 1$ by requiring that the ratio $|\mathbf{x}_{k+1} - \mathbf{x}_{\min}|/|\mathbf{x}_k - \mathbf{x}_{\min}|^p$ converge to a constant, but $p > 2$ is seldom considered. (It should be noted that *first-order* and linear convergence are not the same thing, as first order convergent sequences exist which fail to converge linearly and others can be found which converge superlinearly.) If the function to be minimized is (convex) quadratic, the Newton-Raphson method locates the minimum in one iteration. An algorithm which can be shown to minimize a convex quadratic function in a finite number of iterations is said to exhibit *quadratic termination.* Although the practical implications of this property for arbitrary functions remain unclear, since all suitably smooth functions are quadratic near to an isolated minimum as we have seen in Chapter 3, this is a desirable property.

At the present time theoretical notions of convergence and rate of convergence are principally applicable 'close' to a local minimum. The definition of close is usually in terms of the region of relatively accurate quadratic approximation, but the utility of this definition will of course depend on the nature of the function to be minimized. Even for unimodal functions an algorithm efficient close to a local minimum may be of little use, and functions arising in practice are often not so well behaved. For this reason a minimal requirement for a generally efficient algorithm is *stability*, sometimes called the *down-hill* property. Namely, the sequence of function values at successive iterations is always non-increasing. Stability is relatively easy to ensure in practice, for example by taking suitable precautions when linear searches are employed, or by certain theoretical tests (treated in Section 4.4) when gradient evaluations are permitted.

A concept both more vague and much more difficult to ensure is termed *robustness*. A robust algorithm is one which in practice usually yields the global minimum or a good local minimum of any function of even a large number of variables from a poor initial approximation. In order to avoid premature termination it is important to simultaneously consider termination criteria in terms of relative function decrements, variable increments and, when available, gradient increments. Indeed, a relative function increments criterion alone can cause termination on a plateau, while relative variable increments can cause termination on a very steep slope. One widely used device to avoid premature termination is to perturb the current solution in a random direction within the tolerances of the termination criterion and restart the minimization process. If the process produces no further change the user can have some confidence in having identified a local minimum, since with most algorithms descent is likely to begin again at, for example, a saddle point. Another device is to evaluate the function at a small perturbation on either side of the terminal point along each of the axes in turn until either a lower value is obtained, when the process is restarted, or all n directions are exhausted. A procedure which attempts to avoid termination at a local minimum is to begin the minimization process from several different starting points. Other devices are discussed by Goldstein and Price (1971), Anderssen (1972) and McCormick (1972).

Since an iteration is an arbitrary concept for many algorithms, a practical measure of speed of convergence is the number of function evaluations required. When first or second derivatives are evaluated an equivalent number of function evaluations required to estimate them is often used in comparing algorithms, but this can be misleading. Since the number of function evaluations neglects housekeeping operations such as linear equation solution, etc., a further measure of efficiency is computer time to termination. This is of course actual execution time and should exclude manipulation of input arrays for production codes, multiple core input-output time on time-sharing systems, and time to execute the complex print statements often used in testing. A device for relating results on different computer systems is to consider execution times in terms of the execution time of a standard program.

A robust algorithm with an acceptable practical rate of convergence on all iterations may be termed *reliable*. In spite of theoretical clumsiness and difficulties in specifying criteria for the change-over points, a standard operational practice to improve reliability when function derivatives are available (or can be effectively estimated) sequences methods for use in different neighbourhoods of the global minimum. Since progress of the steepest descent method is satisfactory far from the minimum, it is used to locate the area of a local minimum. When reduction of the function or increment falls below a set tolerance, a direct search or second order method, or both in turn, with super-linear convergence is applied. Alternatively, some of the more recent methods attempt to produce this type of behaviour automatically and there is some evidence that they are consistently better than steepest descent even far from the minimum. We now turn to an introduction to these methods.

4.3 Advanced search methods

The essential difference between advanced search methods and the basic methods described so far is that the directions of search, rather than being kept fixed, are changed to improve the search efficiency. As well as demonstrating the characteristics of an advanced search technique, the two methods described are also considered to be among the most successful of the wealth of search methods at present available.

A development of the one at a time search is to search along each of n mutually *conjugate* directions instead of along the axes directions. The minimum of a quadratic function can then be located exactly in n iterations consisting principally of n linear searches.

The pattern search can also be developed further. The method of *rotating co-ordinates*, which is especially suitable for narrow curved valleys, is similar to the pattern search but maintains a co-ordinate system, and therefore directions of search, such that one axis is aligned with the valley.

Conjugate directions

A quadratic function in n dimensions can be written in the form

$$f(\mathbf{x}) = \tfrac{1}{2}\mathbf{x}^{\mathrm{T}}\mathbf{A}\,\mathbf{x} + \mathbf{b}^{\mathrm{T}}\mathbf{x} + c \qquad (4.1)$$

The n by n matrix $\mathbf{A}$ is an array of coefficients a_{ij}, so that expansion of $\mathbf{x}^{\mathrm{T}}\mathbf{A}\,\mathbf{x}$ gives terms of the type $\frac{1}{2}(a_{ij} + a_{ji})x_i x_j$. The column vector $\mathbf{b}$ contains coefficients of the type $b_k x_k$ and c is a constant. Without loss of generality the $\mathbf{A}$ matrix can be assumed symmetric, i.e. $a_{ij} = a_{ji}$ and $\mathbf{A}^{\mathrm{T}} = \mathbf{A}$. Matrix theory involving eigenvalues, eigenvectors, and the diagonalization of symmetric matrices can then be applied, see for example Birkhoff and Maclane (1953), Chapters 7 to 10.

If f is to have a minimum it must be convex. Since it is symmetric, the matrix $\mathbf{A}$ must therefore be non-negative definite, and if it is of rank n, positive definite. A set of n column vectors $\mathbf{u}_1, \mathbf{u}_2, \ldots, \mathbf{u}_n$ can be defined that are mutually *conjugate with respect to* $\mathbf{A}$, i.e.

$$\mathbf{u}_i^{\mathrm{T}}\mathbf{A}\,\mathbf{u}_j = 0 \quad \text{for } i \neq j. \qquad (4.2)$$

Arranging these vectors as the columns of an $n \times n$ matrix $\mathbf{U}$, it follows for positive definite $\mathbf{A}$ that $\mathbf{U}^{\mathrm{T}}\mathbf{A}\mathbf{U}$ is a diagonal matrix of positive entries and the matrix $\mathbf{A}$ is said to be *diagonalized* by $\mathbf{U}$. If the vectors $\mathbf{u}_i$ are the *eigenvectors* of $\mathbf{A}$, i.e. they satisfy $\mathbf{A}\,\mathbf{u}_i = \mu_i\mathbf{u}_i$, where μ_i, $i = 1, \ldots, n$, are the *eigenvalues* of $\mathbf{A}$, it follows from equation 4.2 that $(\mu_i - \mu_j)\mathbf{u}_i^{\mathrm{T}}\,\mathbf{u}_j = 0$, so that unless $\mu_i = \mu_j$, $\mathbf{u}_i^{\mathrm{T}}\mathbf{u}_j = \mathbf{u}_i \cdot \mathbf{u}_j = 0$.

When $\mu_i = \mu_j$ the corresponding eigenvectors may be chosen orthogonal. The *conjugate directions* defined by the vectors $\mathbf{u}_i$ therefore constitute an orthogonal co-ordinate frame for n dimensional space. If $\mathbf{A}$ is singular and therefore non-negative definite, it has one or more zero eigenvalues. However an orthogonal set of n linearly independent eigenvectors can still be found. There are many sets of conjugate directions defined by n linearly independent vectors $\mathbf{u}_i, \ldots, \mathbf{u}_n$ mutually conjugate with respect to A, but only those defined by the eigenvectors of A are mutually orthogonal. Optimization methods based on sets of conjugate directions search each direction in turn for the minimum.

If the first estimate of the minimum is given by $\mathbf{x}_0$ then the minimum is assumed to be given by a linear combination of the vectors $\mathbf{u}_1, \ldots, \mathbf{u}_n$ and $\mathbf{x}_0$, i.e.

$$\mathbf{x}_{\mathrm{min}} = \mathbf{x}_0 + \Sigma_{i=1}^{n} \lambda_i \mathbf{u}_i \,. \tag{4.3}$$

Therefore the values of the λ_i must be chosen to minimize

$$E = f(\mathbf{x}_0 + \Sigma_{i=1}^{n} \lambda_i \mathbf{u}_i). \tag{4.4}$$

Substituting for f from equation 4.1, expanding the summation, and using equation 4.2 gives

$$f(\mathbf{x}_0 + \Sigma_{i=1}^{n} \lambda_i \mathbf{u}_i) = f(\mathbf{x}_0) + \Sigma_{i=1}^{n} [\tfrac{1}{2} \lambda_i^2 \mathbf{u}_i^{\mathrm{T}} \mathbf{A} \mathbf{u}_i + \lambda_i \mathbf{u}_i (\mathbf{A}\mathbf{x}_0 + \mathbf{b})] \,. \tag{4.5}$$

In the direction of $\mathbf{u}_i$ the search for the value λ_i minimizes a term containing $\mathbf{u}_i$ and λ_i in equation 4.5 which is independent of the λ_j for the other directions. The minimum value of E can therefore be determined by a single linear search in each of the conjugate directions. When the conjugate directions are defined by the eigenvectors of $\mathbf{A}$, this algebraic separability of terms represents the geometric interpretation of the $\mathbf{u}_i$ vectors as a set of orthogonal *principal axes* for the quadratic surface defined by f.

The problem that remains is the development of an efficient procedure for determining n linearly independent conjugate direction vectors. An iterative method for its solution proposed by Powell (1964) is to carry out a succession of 'one at a time' searches in each of n sets of independent directions beginning initially with the co-ordinate directions. If at the kth iteration the n direction vectors are given by $\mathbf{u}_1^k, \mathbf{u}_2^k, \ldots, \mathbf{u}_n^k$ and the current argument of the function is given by $\mathbf{x}_k$, then a linear search in each direction in turn will eventually produce a point $\mathbf{x}_{k+1}$. In the same manner as the one at a time search, each direction is searched from the best point so far determined so that $\mathbf{x}_k$ is progressively updated to $\mathbf{x}_{k+1}$. The direction vector given by $\mathbf{x}_{k+1} - \mathbf{x}_k$ is used to define a new direction of search and the first of the original directions is discarded in order to maintain conjugacy. Hence a new set of directions is defined for the next iteration by

$$\{\mathbf{u}_1^{k+1}, \mathbf{u}_2^{k+1}, \ldots, \mathbf{u}_n^{k+1}\} = \{\mathbf{u}_2^k, \mathbf{u}_3^k, \ldots, \mathbf{u}_n^k, (\mathbf{x}_{k+1} - \mathbf{x}_k)\}. \tag{4.6}$$

A typical search in two dimensions is illustrated in Figure 16. Since contours are elliptical the procedure converges in two iterations. It is obvious from the figure why this method of generating a set of conjugate directions is known as a *parallel subspace technique* (Fletcher, 1969a).

At the initial point no directions are available so the search commences in the axis directions. The first iteration therefore searches direction A to find point 1 and then direction B to find point 2. The iteration is completed by a search in the directions $\mathbf{x}_0$ to point 2, i.e. direction C, to find the point $\mathbf{x}_1$ from which the second iteration is commenced. Before the second iteration the search direction vectors are changed so that directions B and C replace A and B. The second iteration in directions B and C gives progressively point 3, in direction D parallel to B, point 4, in direction E parallel to C, and finally, point $\mathbf{x}_2$, by a search in the new direction F which is in the direction from $\mathbf{x}_1$ to point 4. Since the contours are elliptical, point $\mathbf{x}_2$ is the minimum required. If the process were to be continued, the search directions B and C would be replaced by the directions C and F. In this simple example C and F are the conjugate directions sought. With exact linear minimization this process gives n mutually conjugate directions after at most n iterations.

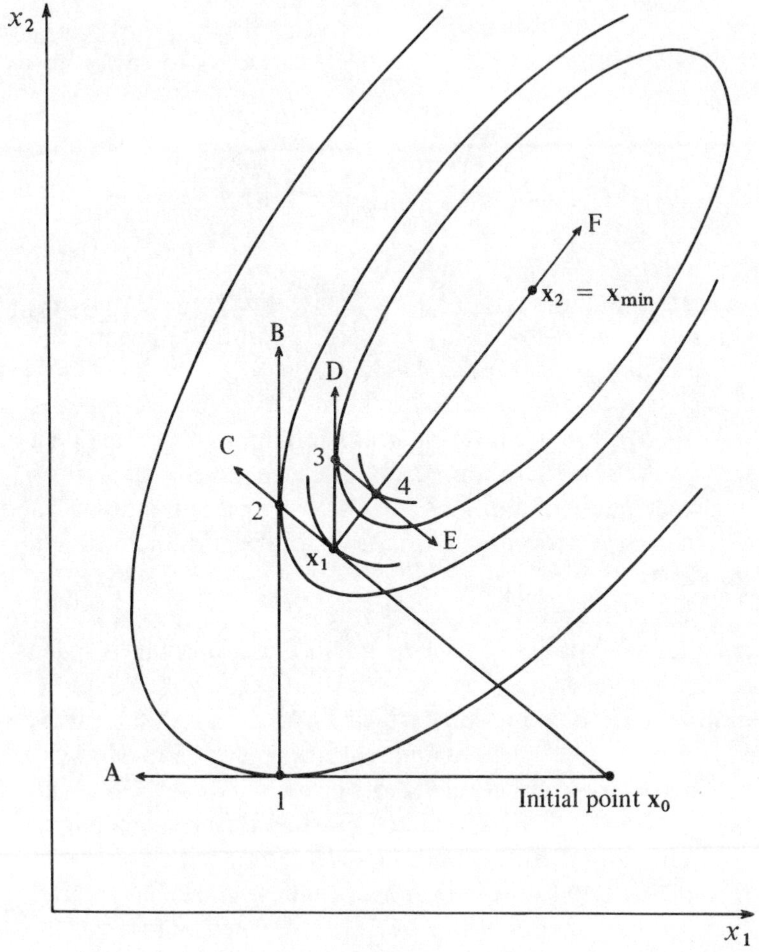

Fig. 16. *Search by the method of conjugate directions.*

To see this, note that if $\mathbf{x}_1$ and $\mathbf{x}_2$ are minima of f in a direction $\mathbf{u}$ from two arbitrary distinct points, then at $\lambda = 0$,

$$\frac{\partial}{\partial \lambda} f(\mathbf{x}_1 + \lambda \mathbf{u}) = 0,$$

i.e.

$$2\lambda \mathbf{u}^{\mathrm{T}} \mathbf{A} \mathbf{u} + \mathbf{u}^{\mathrm{T}} (2\mathbf{A}\mathbf{x}_1 + \mathbf{b}) = 0$$

so that since $\lambda = 0$

$$\mathbf{u}^T(2\mathbf{A}\mathbf{x}_1 + \mathbf{b}) = 0 \tag{4.7}$$

and similarly

$$\mathbf{u}^T(2\mathbf{A}\mathbf{x}_2 + \mathbf{b}) = 0. \tag{4.8}$$

Subtracting 4.7 from 4.8 yields

$$\mathbf{u}^T\mathbf{A}(\mathbf{x}_2 - \mathbf{x}_1) = 0,$$

so that $\mathbf{u}^T$ and $(\mathbf{x}_2 - \mathbf{x}_1)$ are conjugate directions. Applying this result iteratively yields n mutually conjugate linearly independent directions after n iterations and $n(n+1) = n^2 + n$ (exact) linear searches if $\mathbf{A}$ is positive definite and hence of full rank. If $\mathbf{A}$ is of rank less than n the minimum is degenerate and fewer iterations are required. By our earlier result the last n linear searches will be along a set of conjugate directions. There-Therefore the procedure exhibits quadratic termination in n iterations.

In general, however, it is possible for the process to fail to produce a set of linearly independent directions which span n dimensional space. For example, if at the kth iteration the minimum in the direction defined by $\mathbf{u}_1^k$ remains unchanged, $\mathbf{x}_{k+1} - \mathbf{x}_k$ will have no component in this direction. Hence the new set of n direction vectors given by equation 4.6 will fail to span the $\mathbf{u}_1^k$ direction and must be linearly dependent.

Modifications to the method which attempt to overcome this difficulty have been proposed by Powell. He has shown that if the current direction vectors are scaled so that

$$\mathbf{u}_i^T\mathbf{A}\mathbf{u}_i = 1$$

and a square matrix $\mathbf{U}$ is defined consisting of the elements of the current direction vectors arranged as columns, viz.

$$\mathbf{U} = [\mathbf{u}_1\mathbf{u}_2 \ldots \mathbf{u}_n],$$

then the determinant $|\mathbf{U}|$ of $\mathbf{U}$ is maximal if, and only if, the current directions are mutually conjugate. Suppose that at the kth iteration the search direction which gives the greatest reduction in f is observed to be the ith direction. Then it follows that the direction vector $\mathbf{u}_i$ should be removed from

$\mathbf{U}^k$ and the direction $(\mathbf{x}_{k+1} - \mathbf{x}_k)$ used to generate $\mathbf{U}^{k+1}$ from $\mathbf{U}^k$ only if $|\mathbf{U}^{k+1}| > |\mathbf{U}^k|$. Unfortunately a set of n linearly independent conjugate directions may never be obtained in the resulting procedure if one of the directions generated is removed and later replaced or if no new direction is generated because of failure to satisfy the condition imposed.

Zangwill (1968) has given a simplification of Powell's suggestion. He recommends normalizing the $\mathbf{u}_j$ to unit length, i.e. $|\mathbf{u}_j| = \sqrt{\sum_{i=1}^{n} u_i^2} = 1$, and makes use of a variable δ, initially set equal to one, which may be shown by induction to be the determinant of the current $\mathbf{U}$. Suppose λ_k is the maximum of the λ_i's found at the kth iteration (see expression 4.5). Then at the end of the kth iteration if $\lambda_k \delta_k / |\mathbf{x}_{k+1} - \mathbf{x}_k|$ is not less than a prescribed ϵ in the unit interval, δ_{k+1} is set equal to this quantity and the kth column of $\mathbf{U}$ is replaced by $(\mathbf{x}_{k+1} - \mathbf{x}_k)/|\mathbf{x}_{k+1} - \mathbf{x}_k|$. If the test quantity falls short of ϵ the current $\mathbf{U}$ is retained. Thus the columns of the current $\mathbf{U}$ must always span n dimensional space. Although this procedure suffers the same difficulties as Powell's method and no longer exhibits quadratic termination, theoretical convergence to the (global) minimum of a continuously differentiable strictly convex function can be demonstrated.

Zangwill also proposes the following simple modification to the original procedure based on 4.6 (using $\mathbf{u}_i$ vectors normalized to unit length) which precludes the possibility of failing to generate new directions and still terminates quadratically. As the last step of each iteration after the n searches in the directions of the current $\mathbf{U}$ a search is made along a single co-ordinate axis chosen cyclically from the n co-ordinate axes in turn. The result defines $\mathbf{x}_{k+1}$ and the new direction for $\mathbf{U}$. If this $(n + 1)st$ search should result in no change to the current value of $\mathbf{x}$, the function is searched next along the next co-ordinate axis until all n co-ordinate axes are exhausted and the gradient of f must be zero. This procedure not only converges to the minimum of a continuously differentiable strictly convex function, but for quadratic functions it does so in at most n iterations, as required. In terms of function evaluations on test problems Rhead (1971, 1972) reports marginally poorer results with this modification than with Powell's original method. In practice, however, Powell's method sometimes fails to generate new directions close to the minimum, particularly when the number of variables is large. Zangwill's modification overcomes this difficulty. A widely used alternative is to reset the search directions to the co-ordinate directions every $n + 1$ iterations. If the minimum is not located in one such cycle, then f is not locally quadratic. Although then the argument for retaining the current set of directions is weakened, keeping the old directions often works better in practice.

The other principal technique for generating a set of conjugate direction vectors is based on gradient evaluation. Although methods based on this 'projection' technique have not proved as effective as those based on the parallel subspace technique, they will be discussed briefly in the next section.

The theory underlying methods of conjugate directions applies to quadratic functions, and the most effective methods exhibit quadratic termination. Since, however, the most effective procedure for generating conjugate directions is meant to be used with an arbitrary function without knowledg of its derivatives, neither the Hessian matrix nor the gradient vector is explicitly utilized. Recent work by Powell (see for example 1972c) is aimed at generating linearly independent conjugate directions utilizing estimates of the Hessian from function evaluations, but no specific algorithm has so far been proposed. It is natural to expect that methods of conjugate directions should perform well in the region of a minimum where quadratic approximation is fairly accurate. Powell's original method is widely used and has been particularly useful for optimizing likelihood functions in statistics and econometrics where relatively good intial approximations are often available. On the other hand, McCormick and Ritter (1972) have shown that for an arbitrary twice continuously differentiable function, at best a super-linear rate of convergence is achieved by conjugate direction methods at every $n - 1$ or n iterations. (The rate of convergence is in fact $(n - 1)$- or n-step quadratic if the Hessian is suitably continuous.) This suggests that their widespread use may be due as much to their reliability in efficiently locating the region of the global minimum as to their rapid convergence to it. This has certainly been the experience with Powell's method in some applications which involve large numbers of variables.

Rotating co-ordinates

This search method, due to Rosenbrock (1960), incorporates the idea of rotating the co-ordinate system so that the search directions in a search pattern are aligned in such a way that one direction is in the expected direction of the minimum. The remaining directions are then particularly suited to the task of making required corrections to the main search direction as the search progresses.

In practice the co-ordinate system is changed indirectly by the choice of an orthonormal set of search directions, i.e. a set of mutually orthogonal vectors of unit length, within the current co-ordinate system. In common with the method of conjugate directions, the search is started by using the original co-ordinate directions for the first iteration.

Each iteration of the search is divided into two parts.

(i) A pattern type of search is used to determine a suitable direction.

(ii) The search direction vectors are changed to align with the direction vector determined in part (i) of the iteration.

At the kth iteration define a set of n mutually orthonormal direction vectors by $\mathbf{u}_1^k, \mathbf{u}_2^k, \ldots, \mathbf{u}_n^k$ and the respective step lengths $\delta_1^k, \delta_2^k, \ldots, \delta_n^k$. The start of the kth iteration is from the point $\mathbf{x}_k$.

The steps in the search part of each iteration were given by Rosenbrock as follows.

Step: 1. In each search direction i in turn, the function f is evaluated at a point given by the point $\mathbf{x}_k$ plus an incremental step δ_i^k in the ith search direction. The test point is therefore given by $\mathbf{x}_k + \delta_i^k \mathbf{u}_i^k$. If the value of f at the test point does not exceed the value at $\mathbf{x}_k$ the test is successful. The test point then becomes the current $\mathbf{x}_k$ and the value of δ_i^k is multiplied by an expansion factor greater than 1 (usually in the order of 3.0). If the value of f at the test point is greater than the value at $\mathbf{x}_k$, the test is a failure and the value of δ_i^k is multiplied by a negative factor less than 1 (usually -0.5).

2. The cycle of tests is repeated continually until a successful test point is succeeded by a failure in each direction. When this happens the direction involved is not tested further and the next direction is searched.
3. When the search is completed, i.e. all directions have been searched the current point $\mathbf{x}_k$ becomes the point $\mathbf{x}_{k+1}$ from which the iteration $k+1$ is commenced and the direction $\mathbf{x}_{k+1} - \mathbf{x}_k$ is used to define the first direction vector of a new set of orthonormal vectors $\mathbf{u}_1^{k+1}, \ldots, \mathbf{u}_n^{k+1}$.

The second part of the algorithm determines these n unit orthogonal vectors by means of the Gram-Schmidt orthogonalization process, see for example Birkhoff and Maclane, Chapter 7. If the total movement $d_1, d_2, \ldots, d_n$ in each of the search directions is retained, then the vectors $\mathbf{v}_1, \mathbf{v}_2, \ldots, \mathbf{v}_n$ can be defined by

$$\mathbf{v}_n = d_n \mathbf{u}_n \tag{4.7}$$

$$\mathbf{v}_i = d_i\mathbf{u}_i + \mathbf{v}_{i+1}, \quad i = n-1, n-2, \ldots, 1. \tag{4.8}$$

The new set of orthonormal direction vectors are then given by

$$\mathbf{u}_1^{k+1} = \frac{\mathbf{v}_n}{|\mathbf{v}_n|} \tag{4.9}$$

$$\mathbf{u}_i^{k+1} = \frac{\mathbf{w}_i}{|\mathbf{w}_i|} \quad i = 2, 3, \ldots, n \tag{4.10}$$

where

$$\mathbf{w}_i = \mathbf{v}_i - \Sigma_{j=1}^{i-1} (\mathbf{v}_i^T \mathbf{u}_j^{k+1})\, \mathbf{u}_j^{k+1}. \tag{4.11}$$

Several modifications to Rosenbrock's method of rotating

co-ordinates are available which are in general more efficient. The simplest of these, due to Davies, Swann and Campey (Swann, 1964), replaces the pattern type of search by a single variable search in the appropriate directions. If the function is of the type that permits the use of quadratic curve fitting, a considerable reduction in the total number of function evaluations is possible. Further improvements to the orthogonalization procedure are given by Palmer (1969).

Although Rosenbrock's method with linear search does not exhibit quadratic termination, due to the orthonormalization of the search directions in the second part of each iteration and the alignment of one of them towards the minimum, they may tend to converge to the eigenvectors of the Hessian at the optimum. As a result the method can have approximate quadratic termination. On the other hand, it can be as slow as steepest descent on specific functions of two variables. An advantage of rotating co-ordinates is that it tends to follow narrow curving valleys. However, since new directions are used at every search, Powell's technique for cutting down function evaluations when directions are re-searched (Powell, 1964) cannot be used. Powell's method effects linear search by iterated quadratic interpolations. An initial estimate of the second derivative of the function in a given direction is used to allow subsequent prediction of the minimum in that direction from a single new function evaluation and the function value at the current point. In experiments Fletcher (1965) found Powell's method comparable to Rosenbrock's with linear search, but Box (1966) found Powell's method markedly superior. Unfortunately, in recent tests (Himmelblau, 1972), only Rosenbrock's original method was tested. However Powell's method with a standard linear search was found not only by far the best of the direct search methods, but also comparable in efficiency to some of the best advanced gradient algorithms.

4.4 Advanced gradient methods

Advanced gradient algorithms are developments of the basic methods described in Chapter 3. They attempt to guarantee convergence while avoiding the necessity to compute some or all of the derivatives normally required by the basic methods. Most of these are termed *quasi-Newton* methods since they attempt to simulate the Newton-Raphson iteration of Section 3.3 simply from knowledge of the gradient vector of the function to be maintained, often in conjunction with linear searches.

A quadratically terminating method due to Fletcher and Reeves (1964) which makes use of conjugate gradient vectors (with respect to the Hessian matrix) has been utilized in procedures for constrained optimization because of its modest memory requirements. In place of an approximation to the inverse Hessian matrix required by quasi-Newton methods, it only requires two vectors to be stored. The price is a sacrifice of efficiency however and the method is not recommended for general use.

Several modifications of the basic Newton-Raphson method have been proposed for the situation when second derivatives are available. These attempt to take advantage of both the reliable convergence of the method of steepest descent far from the optimum, and the quadratic rate of convergence of the basic second order method close to the optimum. In addition they overcome difficulties caused by a non-positive definite Hessian which can cause the Newton-Raphson method to diverge or stop at a non-stationary point of f (see Powell, 1966). An early application of these ideas was made to the Gauss-Newton iteration of Section 3.3 for the least squares problem, and we shall study this in some detail.

Amongst the most successful quasi-Newton methods are those which fall into the class of algorithms termed *variable metric*. The common features of these methods is that an estimate $\mathbf{S}$ of the inverse Hessian matrix $\mathbf{H}^{-1}$ is continuously updated using information obtained from the changes in the Jacobian gradient vector. Difficulties due to a non-positive definite Hessian are usually avoided by ensuring that $\mathbf{S}$ is maintained positive definite. An effective variant is available in which the required gradients are estimated from function evaluations.

The idea of estimation of derivatives can also be applied to the method of least squares. Since the derivatives needed are of the first order only, as we saw in Section 3.3, a least squares procedure is obtained without the need to compute any derivatives at all.

Conjugate gradients

A technique due originally to Hestenes and Stiefel (1952) for generating a set of conjugate direction vectors from gradient evaluations is based on the easily seen fact that for the quadratic function 4.1,

$$\mathbf{\Delta g} = \mathbf{g}(\mathbf{x} + \mathbf{\Delta x}) - \mathbf{g}(\mathbf{x}) = \mathbf{A}\ \mathbf{\Delta x}. \qquad (4.12)$$

The search directions $\mathbf{u}_1, \ldots, \mathbf{u}_k$, $k \leqslant n$, are arranged to be orthogonal to the set of gradient differences at successive iterations by projecting a set of linearly independent vectors onto the subspace spanned by these differences. Since linear searches are executed, $\mathbf{\Delta x}_k = \lambda_k \mathbf{u}_k$, and using 4.12,

$$\mathbf{u}_k^{\mathrm{T}} \mathbf{A}\, \mathbf{u}_j = \lambda_k^{-1}\, \mathbf{\Delta x}_k^{\mathrm{T}} \mathbf{A}\, \mathbf{u}_j = \lambda_k^{-1}\, \mathbf{\Delta x}_k^{\mathrm{T}}\, \mathbf{\Delta g}_j = 0 \quad j = 1, \ldots, k-1,$$

where $\mathbf{\Delta g}_j = \mathbf{g}(\mathbf{x}_{j+1}) - \mathbf{g}(\mathbf{x}_j)$, as required for conjugacy.

In more detail, suppose $\mathbf{v}_1, \ldots, \mathbf{v}_n$ are a set of linearly independent vectors. Then a set of mutually conjugate directions $\mathbf{u}_1, \ldots, \mathbf{u}_n$ may be constructed by setting $\mathbf{u}_1 = \mathbf{v}_1$ and

$$\mathbf{u}_k = \mathbf{v}_k + \Sigma_{j=1}^{k-1} c_{jk}\, \mathbf{u}_j \quad k = 2, \ldots, n, \qquad (4.13)$$

where the coefficients c_{jk} are chosen to make

$$\mathbf{u}_k^{\mathrm{T}} \mathbf{A}\, \mathbf{u}_j = 0 \quad j = 1, \ldots, k-1 \qquad (4.14)$$

On substitution of 4.13 in 4.14 and simplification it follows that

$$c_{jk} = -\mathbf{v}_k^{\mathrm{T}} \mathbf{A}\, \mathbf{u}_j / \mathbf{u}_j^{\mathrm{T}} \mathbf{A} \mathbf{u}_j \quad j = 1, \ldots, k-1, \quad (4.15)$$

Setting $\mathbf{v}_k = -\mathbf{g}_{k-1} = \mathbf{g}(\mathbf{x}_{k-1})$, $k = 1, \ldots, n$, and using 4.12 yields

$$c_{jk} = -\mathbf{g}_{k-1}^{\mathrm{T}}\, \mathbf{\Delta g}_{j-1} / \mathbf{u}_j^{\mathrm{T}}\, \mathbf{\Delta g}_{j-1} \quad j = 1, \ldots, k-1\,.$$

But after $k-1$ linear searches in $k-1$ conjugate directions we have seen above that 4.1 is minimized in the space spanned by these vectors. It follows by the usual argument that since $\mathbf{x}_{k-1}$ is at a local minimum over this space $\mathbf{g}_{k-1}$ is orthogonal to any search direction in it. Since this space has been constructed from differences of $\mathbf{g}_0, \ldots, \mathbf{g}_{k-2}$,

$$\mathbf{g}_{k-1}^{\mathrm{T}}\mathbf{g}_j = 0 \quad j = 1, \ldots, k-2, \tag{4.16}$$

and c_{jk} vanishes for $j = 0, \ldots, k-2$. Thus successive (conjugate) search directions may be obtained recursively from the formula

$$\mathbf{u}_k = -\mathbf{g}_{k-1} + \frac{\mathbf{g}_{k-1}^{\mathrm{T}}\,\Delta\mathbf{g}_{k-2}}{\mathbf{u}_{k-1}^{\mathrm{T}}\,\Delta\mathbf{g}_{k-2}}\,\mathbf{u}_{k-1} \qquad k = 1, \ldots, n,$$

with $\mathbf{u}_1 = -\mathbf{g}_0$, which may be simplified even further, using 4.16, to

$$\mathbf{u}_k = -\mathbf{g}_{k-1} + \frac{\mathbf{g}_{k-1}^{\mathrm{T}}\,\mathbf{g}_{k-1}}{\mathbf{g}_{k-2}^{\mathrm{T}}\,\mathbf{g}_{k-2}}\,\mathbf{u}_{k-1} \qquad k = 1, \ldots, n, \tag{4.17}$$

requiring storage of only two vectors, a search direction and a gradient, at any point of the kth iteration.

The Fletcher-Reeves (1964) algorithm for use with general functions f generates a new search direction at each iteration according to 4.17. With accurate linear search each new direction is a descent direction. Since it retains a component of the previous direction the resulting algorithm possesses the ridge-aligning property of the method of rotating co-ordinates. For similar reasons to those given above in the discussion of Powell's (1964) method, it has become common practice to reset $\mathbf{u}_k$ to $-\mathbf{g}_{k-1}$ after each cycle of n iterations. In this case, however, resetting is necessary to achieve n-step superlinear convergence.

The Fletcher-Reeves method is the most efficient representative of algorithms, such as the partan method or Powell's (1962) method of conjugate gradients, which generate a set of conjugate directions from gradient evaluations and hence exhibit quadratic termination. (See also Fletcher (1969) or Zoutendijk (1970).) In general, however, methods based on this *projection* technique for constructing conjugate directions, in spite of their use of gradient information, have not proved as effective as methods based on the parallel sub-space technique such as Powell's (1964) method, which requires function evaluation alone.

Modified Newton-Raphson methods

The basic second order increment toward the minimum was given in Section 3.3 by the equation

$$\mathbf{\Delta x} = -\mathbf{H}^{-1}\mathbf{g}. \tag{4.18}$$

An iterative minimization procedure based on this equation can be defined by the iteration.

$$\mathbf{x}_{k+1} = \mathbf{x}_k - \lambda_k \mathbf{H}_k^{-1}\mathbf{g}_k , \tag{4.19}$$

in which a linear search determines the value of λ_k. The predicted step in equation 4.18 is therefore used only indirectly as an indication of direction. In general one cannot accurately find the value of λ_k in 4.19 which minimizes $f(\mathbf{x}_{k+1})$ using only a few values of the objective function. A compromise must be struck between reduction in the value of f and the number of times it must be evaluated. Fletcher (1972) gives an example to show that stopping the search as soon as a lower function value is obtained can be disastrous, while Wolfe (1969) gives some effective methods for fixing λ_k with few function evaluations.

Goldfeld, Quandt and Trotter (1966) propose replacing the Newton-Raphson increment 4.18 by the increment

$$\mathbf{\Delta x} = -[\lambda\mathbf{I} + \mathbf{H}]^{-1}\mathbf{g}, \tag{4.20}$$

where $\mathbf{I}$ is then $n \times n$ identity matrix and λ is adjusted to control the iteration. Since for $\lambda > 1$,

$$[\mathbf{I} + \lambda^{-1}\mathbf{H}]^{-1} = \mathbf{I} - \lambda^{-1}\mathbf{H} + \lambda^{-2}\mathbf{H}^2 \ldots ,$$

as $\lambda \to \infty$, $\mathbf{\Delta x}$ tends to $-\mathbf{g}$, an increment in the steepest descent direction, while if $\lambda \to 0$, $\mathbf{\Delta x}$ tends to the standard second order increment. Although the details were not specified exactly, adjustment of λ from iteration to iteration is made automatically to maintain the rapid convergence of the Newton-Raphson method while taking steps in the steepest descent direction, whenever necessary, to prevent divergence. The parameter λ is maintained large enough to ensure that $\lambda\mathbf{I} + \mathbf{H}$ is positive definite. (We shall return to this idea in a least squares application below.)

Other modified Newton-Raphson methods are available. Goldfeld *et al* (1968) have developed an apparently marginally more reliable but slower variant of their algorithm which replaces the identity matrix **I** by a positive definite matrix which is adjusted at each iteration to bias the increment in the direction of the previous one. Numerical results are given in Goldfeld and Quandt (1972), Chapter 1 and Appendix. The method of Matthews and Davies (1971) replaces **H** by a suitable triangular factorization whenever negative or zero diagonal elements of **H** appear. Some algorithms incorporating similar ideas are compared by Powell (1971a) §3.

Variable metric methods

The slow rate of convergence of the method of steepest descent is due to the possibility that the steepest descent direction $-\mathbf{g}$ and the direction to the minimum may be very nearly perpendicular (cf. Figure 15, § 3.3). The Newton-Raphson direction $-\mathbf{H}^{-1}\mathbf{g}$ may be thought of as involving a second order approximation to the correction to the co-ordinate system required to make the function contours spherical. Then the direction to the minimum and the negative gradient in the new co-ordinates, $-\mathbf{H}^{-1}\mathbf{g}$, will coincide. Thus pre-multiplying $-\mathbf{g}$ by the inverse Hessian is a correction to the *metric* properties of the underlying n-dimensional space of vectors $\mathbf{x}$ which converts the negative gradient to the increment $\mathbf{\Delta x}$. This interpretation of the basic second order increment was proposed by Davidon (1959, 1969) who coined the term *variable metric* (or *variance*) *algorithm* to describe methods which at the kth iteration utilize an increment of the form

$$\mathbf{\Delta x}_k = -\lambda_k \mathbf{S}_k \mathbf{g}_k \qquad (4.20)$$

and up-date the 'metric-correcting transformation' $\mathbf{S}_k$ from iteration to iteration. For efficiency, $\mathbf{S}_k$ should converge to $\mathbf{H}^{-1}$ at the optimum so that such algorithms are quasi-Newton methods. At the kth iteration they require evaluation or approximation of the Jacobian gradient vector $\mathbf{g}_k$ and are completely specified by the updating formula for the $n \times n$ matrix $\mathbf{S}_k$ and the method of determining λ_k.

The most widely used of the variable metric minimization methods is known as the *Davidon-Fletcher-Powell* (DFP) method, as it is a development by Fletcher and Powell (1963) of the variable metric method due to Davidon (1959). In updating the approximation to the inverse Hessian it makes use of the formula

$$\mathbf{S}_{k+1} = \mathbf{S}_k - \frac{\mathbf{S}_k\,\boldsymbol{\Delta}\mathbf{g}_k\,\boldsymbol{\Delta}\mathbf{g}_k^{\mathrm{T}}\,\mathbf{S}_k}{\boldsymbol{\Delta}\mathbf{g}_k^{\mathrm{T}}\,\mathbf{S}_k\,\boldsymbol{\Delta}\mathbf{g}_k} + \frac{\boldsymbol{\Delta}\mathbf{x}_k\,\boldsymbol{\Delta}\mathbf{x}_k^{\mathrm{T}}}{\boldsymbol{\Delta}\mathbf{x}_k^{\mathrm{T}}\,\boldsymbol{\Delta}\mathbf{g}_k}, \qquad (4.21)$$

where $\boldsymbol{\Delta}\mathbf{g}_k = \mathbf{g}_{k+1} - \mathbf{g}_k$ and $\boldsymbol{\Delta}\mathbf{x}_k$ is given by 4.20, and it fixes the value of λ_k with a linear search. The $n \times n$ matrix $\mathbf{S}_0$ can be an arbitrary positive definite symmetric matrix but is usually taken to be the identity matrix $\mathbf{I}$, so that the first increment is in the steepest descent direction $-\mathbf{g}$. The recursion formula 4.21 is known as a *rank-two* (updating) formula since the rank of the matrix correction to $\mathbf{S}_k$ is 2. One linearly independent direction, $\mathbf{S}_k\boldsymbol{\Delta}\mathbf{g}_k$ and $\boldsymbol{\Delta}\mathbf{x}_k$ respectively, is contributed by each term. Note that the expressions in the denominators of these terms are scalar.

Fletcher and Powell proved for a general function f that $\mathbf{S}_k$ positive definite implies $\mathbf{S}_{k+1}$ positive definite. For the quadratic function given by 4.1 they showed that the successive increments $\boldsymbol{\Delta}\mathbf{x}_0, \ldots, \boldsymbol{\Delta}\mathbf{x}_{n-1}$ are a set of conjugate directions and $\mathbf{S}_{n-1}^{-1} = \mathbf{A}$, so that the DFP algorithm exhibits quadratic termination in n steps. The key to these results is the *hereditary* property

$$\mathbf{S}_k\,\boldsymbol{\Delta}\mathbf{g}_j = \boldsymbol{\Delta}\mathbf{x}_j \qquad j = 0, \ldots, k-1 \qquad (4.22)$$

satisfied by the sequence of matrices generated by 4.21. It follows by induction that successive search directions are mutually conjugate with respect to $\mathbf{A}$. We must show that for $k = 1, \ldots, n-1$, $\Delta\mathbf{x}_{k-1}$ is conjugate to $\Delta\mathbf{x}_0, \ldots, \Delta\mathbf{x}_{k-2}$, i.e.

$$\Delta\mathbf{x}_j^{\mathrm{T}} \mathbf{A} \, \Delta\mathbf{x}_k = 0 \qquad j = 0, \ldots, k-1.$$

But for quadratic functions (as we have seen),

$$\Delta\mathbf{g} = \mathbf{A} \, \Delta\mathbf{x} , \tag{4.23}$$

so that, using 4.21, 4.22 and the symmetry of $\mathbf{S}_k$,

$$\Delta\mathbf{x}_j^{\mathrm{T}} \mathbf{A} \, \Delta\mathbf{x}_k = \Delta\mathbf{g}_j^{\mathrm{T}} \, \Delta\mathbf{x}_k = -\lambda_k \, \Delta\mathbf{g}_j^{\mathrm{T}} \mathbf{S}_k \, \mathbf{g}_k = -\lambda_k \, \Delta\mathbf{x}_j^{\mathrm{T}} \, \mathbf{g}_k = 0 ,$$

as required. The orthogonality of $\Delta\mathbf{x}_j$ and $\mathbf{g}_k$ for $j = 0, \ldots, k-1$ follows as usual from the fact that $\mathbf{x}_k$ is a minimum in the subspace spanned by these directions. Since a set of n conjugate directions are linearly independent, we have from 4.22, using 4.23, that

$$\mathbf{S}_n \, \mathbf{A} \, \Delta\mathbf{x}_j = \Delta\mathbf{x}_j \qquad j = 0, \ldots, n-1,$$

i.e. $\mathbf{S}_n \, \mathbf{A} = \mathbf{I}$ or $\mathbf{S}_n^{-1} = \mathbf{A}$, as required.

From the positive definiteness of the $\mathbf{S}_k$ for a general function f it follows that the search direction is never orthogonal to the gradient direction, i.e. $-\mathbf{g}_k^{\mathrm{T}} \mathbf{S}_k \, \mathbf{g}_k \neq 0$, so that the denominators of the correction terms in 4.21 are non-zero and in theory a decrement in the value of f is obtained at each step, i.e. the algorithm is stable. In practice however if f is not unimodal along the line of search (as for example for functions generating narrow curving valleys) and the linear search procedure only finds a local minimum, an increase in the function value may result. Further, round-off errors in the computations of the up-dating process can lead to non-positive definite $\mathbf{S}_k$ and divergence. More recent work of Powell (1971) has shown that $\mathbf{g}_k^{\mathrm{T}} \mathbf{S}_k \, \mathbf{g}_k$ is monotolically decreasing, so that if $\mathbf{g}_k$ is small at any stage – for example when passing close to a saddle-point or some other stationary point of f – then $\mathbf{g}_k^{\mathrm{T}} \mathbf{S}_k \, \mathbf{g}_k$ is forced to be small thereafter. Similar behaviour results with highly asymmetrical functions whose Hessians have large eigen-values leading to near-singularity of $\mathbf{S}$. (Such functions often arise with the sequential unconstrained techniques for constrained optimization discussed in Chapter 5.) As a result, the denominators in 4.21 become very small and cause either overflow in up-dating **S** in the computer, or a very large step due to the near-orthogonality of search and gradient directions. It may therefore be useful to regularly reset **S** to the unit matrix, particularly at the early stages of the minimization. McCormick and Pearson (1969) suggest every $n + 1$ iterations (for the usual reason regarding quadratic termination) and report good computational results with this procedure. It is also important to use multiple termination criteria in terms of the magnitudes of the increment, gradient and function decrement simultaneously. Sargent and Sebastian (1972) present evidence that the DFP algorithm is most efficient when the accuracy of the linear search procedure is rather low.

In an appendix to his report, Davidon (1959) gave a simple symmetric rank-one up-dating formula,

$$\mathbf{S}_{k+1} = \mathbf{S}_k + \frac{(\Delta\mathbf{x}_k - \mathbf{S}_k\ \Delta\mathbf{g}_k)(\Delta\mathbf{x}_k - \mathbf{S}_k\ \Delta\mathbf{g}_k)^{\mathrm{T}}}{\Delta\mathbf{x}_k^{\mathrm{T}}\ \Delta\mathbf{g}_k - \Delta\mathbf{g}_k^{\mathrm{T}}\ \mathbf{S}_k\ \Delta\mathbf{g}_k}, \tag{4.22}$$

and Broyden (1970), Fletcher (1970) and Shanno (1970) (BFS) independently proposed the rank-two- formula

$$\mathbf{S}_{k+1} = \left(\mathbf{I} - \frac{\Delta\mathbf{x}_k\ \Delta\mathbf{g}_k^{\mathrm{T}}}{\Delta\mathbf{x}_k^{\mathrm{T}}\ \Delta\mathbf{g}_k}\right)\mathbf{S}_k\left(\mathbf{I} - \frac{\Delta\mathbf{g}_k\ \Delta\mathbf{x}_k^{\mathrm{T}}}{\Delta\mathbf{g}_k^{\mathrm{T}}\ \Delta\mathbf{x}_k}\right) + \frac{\Delta\mathbf{x}_k\ \Delta\mathbf{x}_k^{\mathrm{T}}}{\Delta\mathbf{x}_k^{\mathrm{T}}\ \Delta\mathbf{g}_k}. \tag{4.23}$$

If precautions are taken to maintain **S** positive definite in 4.22 – by introducing and suitably setting at each iteration a scalar multiplicative parameter in the correction term – then both formulae can be shown to possess the hereditary property 4.22. Hence, when used in lieu of 4.21 with linear searches to set the value of λ_k in the increment 4.20, they share with the DFP algorithm the property of quadratic termination in n steps.

Broyden (1967) first generalized the theoretical properties of the DFP algorithm to a class of variable metric methods generated by a class of symmetric up-dating formulae possessing the hereditary property and utilizing linear search. Fletcher (1970) observed that this class could be generated by taking a convex mixture of the DFP and BFS rank-two up-dating formulae 4.21 and 4.23 as

$$\mathbf{S}_k = (1-\alpha)\,\mathbf{S}_k^{DFP} + \alpha\,\mathbf{S}_k^{BFS} \qquad 0 \leqslant \alpha \leqslant 1. \quad (4.24)$$

He deduced that $\mathbf{S}^{BFS}$ exceeds $\mathbf{S}^{DFP}$ in the sub-space spanned by $\mathbf{S}_k\,\boldsymbol{\Delta}\mathbf{g}_k$ and $\mathbf{x}_k$ by a matrix of rank one. Huang (1970) proved a number of theoretical results for a very general three parameter class of up-dating formulae used with linear search, i.e. such that $\mathbf{g}_{k+1}^T\,\boldsymbol{\Delta}\mathbf{x}_k = 0$. This class includes non-symmetric matrices $\mathbf{S}$ such as would be obtained as inverse Hessians of functions with discontinuous second derivatives. It can be shown that requiring symmetry and the hereditary property 4.22 restricts the number of free parameters in Huang's class to one. Powell (1972a,b) has parametrized this sub-class as

$$\mathbf{S}_{k+1} = \mathbf{S}_{k+1}^{BFS} + \beta\,\mathbf{w}\,\mathbf{w}^T\,, \quad (4.25)$$

where the vector $\mathbf{w}$ is given by

$$\mathbf{w} = \mathbf{S}_k\,\boldsymbol{\Delta}\mathbf{g}_k - \frac{\boldsymbol{\Delta}\mathbf{g}_k^T\,\mathbf{S}_k\,\boldsymbol{\Delta}\mathbf{g}_k}{\boldsymbol{\Delta}\mathbf{g}_k^T\,\boldsymbol{\Delta}\mathbf{x}_k}\,\boldsymbol{\Delta}\mathbf{x}_k\,, \quad (4.26)$$

which includes the DFP, Davidon rank-one and BFS up-dating formula.

Huang demonstrated that for a quadratic function f the sequence of points $\mathbf{x}_1, \mathbf{x}_2, \ldots$ generated by algorithms up-dating $\mathbf{S}$ according to 4.25 depends only on $\mathbf{x}_0$ and $\mathbf{S}_0$ and not on β, providing that λ is set by linear search at each iteration and the value

$$\beta = -1/\mathbf{g}_{k+1}^T\,\mathbf{S}_k\,\mathbf{g}_{k+1}$$

is never used. (In the quadratic case the sequence is finite, with a number terms not exceeding n.) Dixon (1972), proved the remarkable theorem that this result holds for general f for which the *level set* $\{\mathbf{x} : f(\mathbf{x}) \leqslant f(\mathbf{x}_0)\}$ is closed and bounded and any ambiguity in the definition of λ_k due to multiple optima along a line is resolved in a consistent way. It follows that any variation in performance of up-dating formulae with linear search must be attributable to numerical accuracy in the up-dating formula and its sensitivity to accuracy in locating the minimum along a line. With a view to overcoming these difficulties, Oren and Luenberger (1974) have begun the investigation of automatic rescaling of f in a class of self-scaling variable metric algorithms equivalent to Huang's class.

Adding to closed and boundedness conditions on the level set the condition that f possesses second derivatives bounded in the sense that the modulus of the Hessian as an n^2 vector, $|\mathbf{H}(\mathbf{x})|$, is bounded on this set, Powell (1972) demonstrated the theoretical convergence of algorithms of this class to a stationary point of f, i.e. where $\mathbf{g} = \mathbf{o}$, for convex functions. McCormick (1969) has proved that if the DFP algorithm with resetting every $n + 1$ iterations is applied to a general function f under these conditions, then all the cluster points of the sequence $\mathbf{x}_q$, $q = 0, 1, \ldots$, of reset points are stationary points of f. Moreover, if $\mathbf{x}_q$ converges to a local minimum $\mathbf{x}_{\min}$ at which $\mathbf{H}$ is positive definite and satisfies the *Lipschitz condition*

$$|\mathbf{H}(\mathbf{x}) - \mathbf{H}(\mathbf{x}_{\min})| \leqslant c\,|\mathbf{x} - \mathbf{x}_{\min}| \qquad (4.27)$$

for some $c > 0$ over the level set, then convergence is quadratic. That is to say the convergence of the sequence of points $\mathbf{x}_0, \mathbf{x}_1, \ldots$ generated by the algorithm is $(n + 1)$-step quadratic. Powell (1971) has shown that in this situation the DFP algorithm without resetting converges super-linearly.

Both the rank-one and the BFS rank-two matrix up-dating formulae 4.22 and 4.23 have been proposed for use with algorithms which set the step length in the increment $\mathbf{\Delta x}_k = -\lambda_k \mathbf{S}_k \mathbf{g}_k$ without recourse to linear search. If for some $0 < \epsilon < 1$ a stability criterion in terms of the first order approximation to the decrement given by

$$f(\mathbf{x}_k) - f(\mathbf{x}_{k+1}) \geqslant \epsilon\,|\mathbf{g}_k^{\mathrm{T}}\,\mathbf{\Delta x}_k| \qquad (4.28)$$

is enforced at each iteration instead of minimization of f in the search direction, many fewer evaluations may be required. It can be shown that the rank-one formula 4.22 is the only formula with a correction in the space spanned by $\mathbf{S}_k\,\mathbf{\Delta g}_k$ and $\mathbf{\Delta x}_k$ which yields quadratic termination without linear search. The difficulty in using it is that the correction in 4.22 may be unbounded, and if $\mathbf{x}_{k+1} = \mathbf{x}_k - \mathbf{S}_k\,\mathbf{g}_k$ happens to minimize f in the direction $-\mathbf{S}_k\,\mathbf{g}_k$, $\mathbf{S}_{k+1}$ is singular or undetermined.

All three updating formulae 4.21, 4.22 and 4.23 fail to maintain positive definiteness of **S** without linear search, but a necessary and sufficient condition for maintenance of positive definiteness with the DFP and BFS formulae is that

$$\mathbf{\Delta x}^{\mathrm{T}}\ \mathbf{\Delta g} > 0.$$

If this condition is imposed at each iteration (possibly by setting $\lambda_k > 1$), then Fletcher (1970) has shown that the rank-one formula is given by 4.24 with

$$\alpha = \mathbf{\Delta x}_k^{\mathrm{T}}\ \mathbf{\Delta g}_k / (\mathbf{\Delta x}_k^{\mathrm{T}}\ \mathbf{\Delta g}_k - \mathbf{\Delta g}_k^{\mathrm{T}}\ \mathbf{S}_k\ \mathbf{\Delta g}_k)$$

not in the unit interval. He demonstrates that for the quadratic function 4.1 all the up-dating formulae given by 4.24 with $0 \leqslant \alpha \leqslant 1$ have the property that $\mathbf{S}_k$ tends to $\mathbf{A}^{-1}$ in a suitable sense, but formulae with α outside the unit interval do not. He therefore proposes an algorithm in which $\lambda_k = 1, 10^{-1}, 10^{-2} \ldots$ until 4.27 is satisfied, and **S** is up-dated using the BFS or DFP formula according as $\mathbf{\Delta x}^{\mathrm{T}}\mathbf{\Delta g}$ is greater than or equal to, or strictly less than, $\mathbf{\Delta g}^{\mathrm{T}}\,\mathbf{S}\,\mathbf{\Delta g}$. In practice most steps are effected with $\lambda = 1$ after the intial iterations. Although there is evidence that a similar algorithm based solely on the BFS up-dating formulae would be equally successful (see Sargent and Sebastion, 1972, and Dixon, 1972a), Himmelblau (1972) found Fletcher's algorithm an order of magnitude superior to the DFP algorithm and the rank-one algorithm of Broyden (1965), which were the best of the earlier methods. (In both these methods linear searches were effected using an iteration of the Davies, Swann and Campey method followed by iterative quadratic fitting as discussed in Section 2.1.) Dixon reports efficiency comparable to Fletcher's algorithm for the BFS up-dating formula used along with a procedure for bracketing the minimum along the line of search by iterative quadratic fitting until a minimal reduction in f is obtained.

Murtagh and Sargent (1970) consider algorithms using an up-dating formula in which no zero divisors occur and making increments $\mathbf{\Delta x}_k$ in a down-hill direction so as to satisfy the *angle test*

$$|\mathbf{g}_k \, \boldsymbol{\Delta}\mathbf{x}_k| \geqslant \delta \, |\mathbf{g}_k| \, |\boldsymbol{\Delta}\mathbf{x}_k| \tag{4.29}$$

for some $\delta > 0$. They show that under the conditions on f specified in Dixon's result above, a sequence of points $\mathbf{x}_0, \mathbf{x}_1, \mathbf{x}_2, \ldots$ is generated which satisfies 4.27 for some $\epsilon > 0$ and converges to a stationary point of f. With the precautions mentioned in Section 4.1 this will be a local minimum $\mathbf{x}_{\min}$. McCormick and Ritter (1972) prove a result which shows that convergence is super-linear if, in addition, $|\mathbf{S}_k - \mathbf{H}_k^{-1}| \to 0$ as $k \to \infty$. When the Hessian $\mathbf{H}$ satisfies the Lipschitz condition 4.27 and for some $\vartheta \leqslant n$,

$$|\boldsymbol{\Delta}\mathbf{x}_k - \mathbf{H}_k^{-1} \mathbf{g}_k| / |\mathbf{g}_k| \, |\mathbf{g}_{k-\vartheta}| \to 1 \text{ as } k \to \infty,$$

they show that convergence is $(\vartheta + 1)$-step quadratic. These properties, and the value of ϑ for the DFP algorithm and the other algorithms of the class defined by 4.25 and 4.26, are not as yet established. However Powell (1970) proposes a quasi-Newton algorithm utilizing an updating formula in this class and an increment determined by reducing a quadratic approximation to f in the 2-dimensional space spanned by $\mathbf{g}$ and $\mathbf{Sg}$ for which it can be shown that $|\mathbf{S}_k - \mathbf{H}_k^{-1}| \to 0$, and hence that convergence is superlinear. His (1971) paper gives a similar result for the DFP algorithm.

In the situation in which gradient information is not available, Stewart (1967) provides a version of the DFP algorithm with linear search which uses forward difference approximations such as

$$\frac{f(\mathbf{x} + h_i \mathbf{l}_i) - f(\mathbf{x})}{h_i} \approx \frac{\partial f}{\partial x_i}(x) \tag{4.30}$$

to estimate the components of the gradient vector $\mathbf{g}$. If the perturbation h_i is too small, cancellation error is high and error in the $\mathbf{S}$ matrix updating formula 4.21 will be untenable. On the other hand, if h_i is too large the estimation error in 4.30 is excessive. With these considerations in mind Stewart's algorithm fixes h_i in terms of the curvature of f in the ith co-ordinate direction $\mathbf{l}_i$; sometimes replacing 4.30 by the central difference approximation

$$\frac{f(\mathbf{x}+h_i\mathbf{l}_i)-f(\mathbf{x}-h_i\mathbf{l}_i)}{2h_i} \tag{4.31}$$

to control error. The effect of his procedures is to reduce the perturbations in the difference formulas as a minimum is approached. However Gill and Murray (1971) argue in favour of a constant perturbation and give a rule for changing from 4.30 to 4.31 based on the lower limit on h_i required to achieve a function decrement.

Lill (1970) has developed a version of Stewart's algorithm with modified linear search which gives good results. However Himmelblau (1972) found Stewart's algorithm with golden section linear search, although nearly comparable to the DFP method with the same search, inferior to Powell's (1964) method with linear search effected by an iteration of the Davies, Swann and Campey method followed by iterative quadratic fitting.

When difference approximations are used with variable metric algorithms without linear search, care must be taken to prevent false indications in descent and angle tests made according to 4.29 and 4.28, and in termination criteria. In particular, error in the difference approximation may cause run-on even when a local minimum has been accurately located.

If $\mathbf{S}$ is initially chosen to be the unit matrix in variable metric methods, the first iteration of the minimization process is identical to steepest descent. It has already been noted that rapid convergence is obtained from the steepest descent method at points removed from the minimum. As the approximation to $\mathbf{H}^{-1}$ becomes more exact, the minimization process progressively changes to a quadratically convergent second order method which is particularly suitable as the minimum is approached. It is therefore to be expected that variable metric methods should be highly effective and in practice this expectation has been borne out with functions of up to 200 variables. Current opinion favours the view that these methods used with difference approximations based on function evaluations are also the best approach when analytic derivatives are not available. Surveys by Powell (1971a, 1972c) Dixon

(1972a) and Sargent and Sebastion (1972) give current research developments and numerical experience in more detail.

Modified least squares methods

As described in Section 3.3, the least squares method minimizes the sum of the squares of m functions $f_1(\mathbf{x}), f_2(\mathbf{x}), \ldots, f_m(\mathbf{x})$ of the n parameters of $\mathbf{x}$, where $m > n$. It makes use of the fundamental second order increment toward the minimum given by

$$\Delta\mathbf{x} = -\mathbf{H}^{-1}\mathbf{g}.$$

The Jacobian gradient vector $\mathbf{g}$ and the Hessian matrix $\mathbf{H}$ are then expressed in terms of the vector

$$\mathbf{f} = [f_1(\mathbf{x})\, f_2(\mathbf{x}) \ldots f_m(\mathbf{x})]^{\mathrm{T}}$$

and a matrix of partial differential coefficients

$$\mathbf{A} = \begin{bmatrix} \dfrac{\partial f_1}{\partial x_1} & \dfrac{\partial f_1}{\partial x_2} & \cdots & \dfrac{\partial f_1}{\partial x_n} \\ \vdots & & & \vdots \\ \dfrac{\partial f_m}{\partial x_1} & \cdots & \cdots & \dfrac{\partial f_m}{\partial x_n} \end{bmatrix}.$$

In Section 3.3 it was shown for a least squares increment that $\mathbf{g}$ and a first order approximation to $\mathbf{H}$ are given by

$$\mathbf{g} = 2\,\mathbf{A}^{\mathrm{T}}\mathbf{f} \tag{4.32}$$

$$\mathbf{H} = 2\,\mathbf{A}^{\mathrm{T}}\mathbf{A}, \tag{4.33}$$

and the increment toward the minimum becomes

$$\Delta\mathbf{x} = -[\mathbf{A}^{\mathrm{T}}\mathbf{A}]^{-1}\mathbf{A}^{\mathrm{T}}\mathbf{f}\,. \tag{4.34}$$

The minimization strategy involved in variable metric methods takes advantage of the reliable early convergence of steepest descent far from the optimum, while automatically switching to a rapidly convergent approximation to the Newton-Raphson iteration as the minimum is approached. This strategy was proposed for the least squares solution of non-linear equations and curve-fitting problems a number of years ago by Levenberg (1944). The idea was to replace the Gauss-Newton least squares increment 4.34 by the increment

$$\mathbf{\Delta x} = -[\lambda \mathbf{I} + \mathbf{A}^T \mathbf{A}]^{-1} \mathbf{A}^T \mathbf{f}, \tag{4.35}$$

where λ may be adjusted to control the iteration and the $n \times n$ identity matrix **I** may be replaced by a positive diagonal matrix of squared scaling factors to scale the variables (see Marquardt, 1963). As $\lambda \to \infty$, $\mathbf{\Delta x}$ tends to $\mathbf{A}^T \mathbf{F}/\lambda$, an increment in the steepest descent direction, while if $\lambda \to 0$, $\mathbf{\Delta x}$ tends to the standard Gauss-Newton increment. An attempt was to be made to choose λ at each iteration so as to maintain the rapid convergence of the original method to as large an extent as possible, while taking steps in the steepest descent direction when necessary to prevent divergence. Levenberg proposed a time-consuming linear search for the optimum value of λ at each iteration, but a simple scheme devised by Marquardt for choosing λ at each iteration has been widely used and found to be reasonably efficient. From an arbitrary initial value of λ Marquardt proposed dividing or multiplying λ by powers of a factor ϑ (10 was suggested) until the total error sum of squares

$$E = f(\mathbf{x} + \lambda \mathbf{\Delta x}) = \mathbf{f}(\mathbf{x} + \lambda\, \mathbf{\Delta x})^T\, \mathbf{f}(\mathbf{x} + \lambda\, \mathbf{\Delta x})$$

is reduced. On many problems an average of only slightly more than one solution of the linear system

$$[\lambda \mathbf{I} + \mathbf{A}^T \mathbf{A}]\ \mathbf{\Delta x} = -\mathbf{A}^T \mathbf{f},$$

and the corresponding evaluation E, is required.

However Fletcher (1971) points out some difficulties with Marquardt's method. An obvious one is that a poor initial choice of λ may require several evaluations of the $\mathbf{\Delta x}$ and E before the sum of squares is reduced. Further, the initial reduction of λ by a factor such as 10 may lead to too small an increment Δx with the effect that an inefficient average number of evaluations of E near 2 may result. Similarly, Marquardt's geometric scheme for reducing λ can have the effect that after a sequence of successful iterations followed by a failure to reduce E, a number of increases of λ, with consequent error evaluations, may be necessary before further progress is made. Finally, when using least squares to solve non-linear equation systems, so that the optimum value of total error E is zero, the Gauss-Newton method converges quadratically, while at best a super-linear rate of convergence is possible with Marquardt's method due to the geometric reduction of λ.

Fletcher has proposed and tested modifications to Marquardt's algorithm designed to overcome these difficulties. To obviate the arbitrary initial choice and reduction of λ, Fletcher's method computes the Marquardt increment $\mathbf{\Delta x}$ given by 4.35 with the current value of λ ($\lambda = 0$ at the first iteration). It then compares the resulting reduction in the total error sum of squares $f = \mathbf{f}^T\mathbf{f}$ to that predicted by the first order approximation $\mathbf{f} + \mathbf{A}\,\mathbf{\Delta x}$ to $\mathbf{f}$ at $\mathbf{x} + \mathbf{\Delta x}$, namely

$$-2\,\mathbf{f}^T\,\mathbf{A}\,\mathbf{\Delta x} - \mathbf{\Delta x}^T\,\mathbf{A}^T\,\mathbf{A}\,\mathbf{\Delta x}\,. \qquad (4.36)$$

If the ratio R of the actual reduction to the predicted linear reduction 4.36 is close to one, then the first order approximation to the Hessian of the total error sum of squares involved in the Gauss-Newton increment 4.34 (see 3.30) is valid and λ should be reduced for the next iteration. On the other hand if R is near zero or negative, a more nearly steepest descent iteration should be executed and thus λ should be increased. In experiments it was found that an acceptable strategy is to reduce λ if R exceeds 0.75, to increase λ if R falls below 0.25, and to leave λ unchanged for the next iteration if $0.25 \leqslant \lambda \leqslant 0.75$.

To overcome the disadvantages of Marquardt's geometric scheme for reducing λ, Fletcher's method halves λ unless it falls below a cut-off level λ_c equal to the reciprocal of either the largest element or the trace, i.e. the sum of the diagonal elements, of $(\mathbf{A}^T\,\mathbf{A})^{-1}$. When the calculated λ falls below λ_c the Gauss-Newton increment is used next and λ_c is recalculated. (Otherwise at the next iteration the current value of λ_c is maintained.)

When λ is to be increased, a factor between 2 and 10 is used. The argument used to derive expression 3.13 for the eliminating linear search in the method of steepest descent can be applied to the present problem to give the λ approximating the minimum of f in the direction $\mathbf{\Delta x}$ as

$$\alpha = \mathbf{f}^T \mathbf{A}\, \mathbf{\Delta x}\, [2\, \mathbf{f}^T \mathbf{A}\, \mathbf{\Delta x} + f(\mathbf{x}) - f(\mathbf{x} + \mathbf{\Delta x})]^{-1}.$$

Since for large values of λ, increasing λ to $\vartheta\lambda$ approximately changes $\mathbf{\Delta x} \approx \mathbf{A}^T \mathbf{F}/\lambda$ to $\mathbf{\Delta x}/\vartheta \approx \mathbf{A}^T\mathbf{F}/\vartheta\lambda$, α^{-1} is used to multiply λ unless it falls outside the interval 2 to 10, when the appropriate end point is used. After the early iterations the value 2, which approximately halves the current increment, is sufficiently efficient. The advantage of the choice of cut-off level λ_c given above is that, as Fletcher has shown, increasing λ from 0 to approximately λ_c after a Gauss-Newton iteration also approximately halves the current increment.

Fletcher's modifications have proved successful even on functions considered awkward. They lead to an average number of total error evaluations per iteration only slightly over one, and due to the cut-off level yielding Gauss-Newton iterations to quadratic convergence near the optimum when solving non-linear equations. The modified Marquardt method has the disadvantage in complicated statistical and other problems of requiring the direct computation of first derivatives of the individual error functions f_j, $j = 1, \ldots, m$. Brown and Dennis (1970) analyse the performance of the Gauss-Newton and Marquardt-Levenberg algorithms with these derivatives, i.e. the matrix $\mathbf{A}$, calculated from difference approximations, and these could also be used with the modified version of the latter. However in this situation another method is widely used, to which we now turn.

Least squares without derivatives

In the method of *least squares without derivatives* due to Powell (1965) the first derivatives are not computed or approximated by standard difference formulae (see 4.30 and 4.31) directly. Instead they are estimated, and as the minimization proceeds the estimates are improved. Specifically, the basic least squares method is modified by the introduction of n linearly independent directions defined as for the method of conjugate directions by the n vectors $\mathbf{u}_1, \mathbf{u}_2, \ldots, \mathbf{u}_n$. The square matrix $\mathbf{U} = [\mathbf{u}_1\ \mathbf{u}_2 \ldots \mathbf{u}_n]$ is also used. An $m \times n$ matrix $\mathbf{C}$ contains the estimates of the directional derivatives of each of the m functions in each of the n directions $\mathbf{u}_1, \ldots, \mathbf{u}_n$. $\mathbf{C}$ is therefore given by

$$\mathbf{C} = \mathbf{A}\,\mathbf{U}, \tag{4.37}$$

which differs from the true directional derivative because of the second order terms dropped from $\mathbf{A}$ (see the discussion in Section 3.3). Each column $\mathbf{c}_j$ of $\mathbf{C} = (c_{ij})$ is scaled to unit length so that $|\mathbf{c}_j| = 1$, i.e.

$$\Sigma_{i=1}^{m} c_{ij}^2 = 1 \qquad \text{for } j = 1, 2, \ldots, n. \tag{4.38}$$

The direction defined by $\mathbf{\Delta x}$ is related to the n independent directions by a vector $\mathbf{q}$ such that

$$\mathbf{\Delta x} = \mathbf{U}\,\mathbf{q}. \tag{4.39}$$

Powell's least squares method replaces the standard increment given by equation 4.34 by one computed using equation 4.37 in conjunction with the basic equation

$$\mathbf{q} = -(\mathbf{C}^{\mathrm{T}}\mathbf{C})^{-1}\mathbf{C}^{\mathrm{T}}\,\mathbf{f}. \tag{4.40}$$

The iteration is completed by finding the value $\lambda_{\mathbf{min}}$ which minimizes the objective function given by $f(\mathbf{x} + \lambda\mathbf{\Delta x}) = \mathbf{f}^{\mathrm{T}}(\mathbf{x} + \mathbf{\Delta x})\ \mathbf{f}\ (\mathbf{x} + \mathbf{\Delta x})$ by a linear search.

The search procedure used by Powell determines $\lambda_{\mathbf{min}}$ and gives additionally an estimate of the derivative of the vector function $\mathbf{f}$ at $\mathbf{x}$ in the direction $\mathbf{\Delta x}$. The linear search algorithm due to Powell (1964) provides three values of λ equal to λ_1, λ_2, and λ_3. A quadratic curve is fitted to f as a function of λ (as described in Section 2.1) and the minimum point $\lambda_{\mathbf{min}}$ determined. If λ_1 gives the lowest value of f and λ_2 gives the next lowest, then an estimate of the directional derivatives of $\mathbf{f}$ along the line of search is given by the vector

$$\mathbf{v} = \frac{1}{(\lambda_1 - \lambda_2)}[\mathbf{f}(\mathbf{x} + \lambda_1\,\mathbf{\Delta x}) - \mathbf{f}(\mathbf{x} + \lambda_2\,\mathbf{\Delta x})]. \tag{4.41}$$

Since the directional derivatives of $f = \mathbf{f}^T\mathbf{f}$ along the line of search must be zero at λ_{min}, an improved approximation $\mathbf{w}$ of $\mathbf{v}$ can be obtained from

$$\mathbf{w} = \mathbf{v} - \mu\mathbf{f}(\mathbf{x} + \lambda_{min}\boldsymbol{\Delta}\mathbf{x}), \tag{4.42}$$

where

$$\mu = \frac{\mathbf{v}^T \mathbf{f}(\mathbf{x} + \lambda_{min}\boldsymbol{\Delta}\mathbf{x})}{\mathbf{f}^T(\mathbf{x} + \lambda_{min}\boldsymbol{\Delta}\mathbf{x})\,\mathbf{f}(\mathbf{x} + \lambda_{min}\boldsymbol{\Delta}\mathbf{x})}. \tag{4.43}$$

The vectors $\mathbf{w}$ and $\boldsymbol{\Delta}\mathbf{x}$ are now scaled to be unit vectors as in equation 4.38 for the columns of $\mathbf{C}$.

Before the next iteration the jth direction vector $\mathbf{u}_j$ is replaced by $\boldsymbol{\Delta}\mathbf{x}$ and the jth column of $\mathbf{C}$ by $\mathbf{w}$. Here j corresponds to the value of $i = 1, 2, \ldots, n$ which gives the maximum absolute value of the product of the ith column of $\mathbf{Q}$ and the ith element of $\mathbf{C}^T\mathbf{f}$.

For the first iteration the co-ordinate directions are used as the directions $\mathbf{u}_1, \ldots, \mathbf{u}_n$. A first approximation to the function plus suitable increments allows $\mathbf{A}$ to be computed using differences. The scaling factors required by equation 4.26 are placed in order down the diagonal of $\mathbf{U}$ with all off diagonal terms zero. The initial $\mathbf{C}$ matrix is therefore easily obtained.

Powell has shown that $\mathbf{C}^T\mathbf{C}$ converges toward the identity matrix and that $(\mathbf{U}\mathbf{U}^T)^{-1}$ converges toward the true Hessian $\mathbf{H}$ at the minimum. The technique has been found in practice to be comparable to the ordinary least squares method, but it has the advantage of not requiring derivatives. It is particularly recommended for use in solving systems of non-linear simultaneous equations, as described in Section 1.2, when $\mathbf{f}$ may already consist of the gradients of the functions involved.

4.5. Minimax methods

The minimax error criterion defined in Section 1.2 is, like the least squares criterion, used mainly in curve fitting and approximation problems. In these problems a function is specified either as a discrete set of empirical data or mathematically. In either case the difference or *error* between a specified function and an approximating function dependent on several parameters must be evaluated at a number of values of an independent variable t. Choosing the parameters to minimize the maximum error between the two functions at these sampled points tends to produce an approximation to the specified function for which the error oscillates about the required curve as shown in Figure 17. The minimax criterion is particularly appropriate when the approximating function is required for the purpose of interpolation or prediction of the values of a specified function given by a table of calculated or observed values at a grid of t points. The approximation should be easy to calculate (e.g. a polynomial) so that interpolation or prediction of the given function can be effected by evaluating the approximation at the relevant point with the knowledge that the resulting error is within the minimax error.

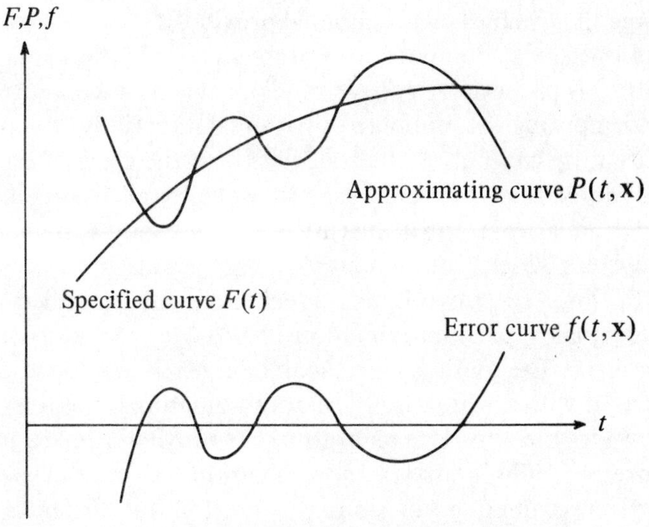

Fig. 17. *Minimax approximation.*

The *equal ripple* approximation specifically requires error peaks of minimum equal amplitude. The number of peaks within a region of approximation may also be specified. The motivation for this form of approximation is the fact, demonstrated below, that when a continuous function is to be approximated by a polynomial of specified degree, the polynomial providing the equal ripple approximation satisfies the minimax error criterion. For certain approximands such as polynomials of high degree, or more generally, sufficiently differentiable functions, the minimax polynomial approximation can be determined or approximated using expansions for the approximand in terms of Taylor, Fourier, Chebyshev, Legendre, etc., series from classical analysis, see Fox and Parker (1968). If the approximand is a polynomial of degree n, the Chebyshev polynomial of degree $n - 1$ is the minimax polynomial of degree $n - 1$. For a general continuous approximand, an $(n - 1)st$ degree polynomial approximation which generates an error function with $n + 1$ alternating maxima and minima sandwiches the minimax polynomial error between the greatest and least of these absolute errors. Since the Chebyshev polynomial provides a thinner sandwich and is easier to compute than other standard polynomials with this property (e.g. Legendre's), a widely used alternative to computing the minimax polynomial directly is to approximate it by computing the coefficients of the Chebyshev series. Polynomials are of course easy to evaluate, so that accurate polynomial approximations are useful in complex numerical processes which require many function evaluations.

In this section two iterative techniques are described which give equal ripple approximations – provided the solutions converge. The methods are capable of generalization and may be used with a wide variety of approximating functions. However the existence and properties of equal ripple and, more generally, minimax approximations for general specified and approximating functions are questions for current mathematical research. They are introduced here in terms of polynomial approximations for the reasons set out in the previous paragraph.

The Remez method

The method of equal ripple approximation to prescribed functions due to Remez (1934) was developed in its basic form for polynomials. A description in English is given by Meinardus (1967), Chapter 7. Consider the continuous function F of an independent variable t. This is to be approximated on a specified interval for t by a polynomial P of degree $n - 1$ such that the error function $f = F - P$ has equal ripple peaks of amplitude E. The problem is then to determine the n coefficients of the polynomial

$$P(t, \mathbf{x}) = \Sigma_{i=1}^{n} x_i t^{i-1} \tag{4.44}$$

where $\mathbf{x} = [x_1 x_2 \ldots x_n]^{\mathrm{T}}$, so as to minimize E.

Provided that the approximand is not a polynomial of degree less than or equal to $n - 1$, when the problem is trivial, there exists a unique $\mathbf{x}_{\min}$ with the property that the resulting error function defined by $f(t, \mathbf{x}_{\min}) = F(t) - P(t, \mathbf{x}_{\min})$ has exactly $n + 1$ extrema in the specified interval whose values are successively $\pm E_{\min}$. These extrema are the $n - 1$ maxima and minima within the interval, plus the non-zero error at its end points. (A proof relying on functional analysis is outlined in Fox and Parker, Chapter 2.) The polynomial corresponding to $\mathbf{x}_{\min}$ satisfies the minimax criterion, for if $\bar{\mathbf{x}}$ generated a polynomial with error $f(t, \bar{\mathbf{x}})$ having a smaller maximum modulus E on the interval of approximation, then

$$\begin{aligned} P(t, \bar{\mathbf{x}}) - P(t, \mathbf{x}_{\min}) &= [F(t) - P(t, \mathbf{x}_{\min})] - [F(t) - P(t, \bar{\mathbf{x}})] \\ &= f(t, \mathbf{x}_{\min}) - f(t, \bar{\mathbf{x}}). \end{aligned}$$

But since $E < E_{\min}$, this polynomial has opposite signs at $n + 1$ successive points in the interval and hence has at least n zeros. Since its degree is at most $n - 1$, it must vanish identically.

The *Remez method* for finding the minimax polynomial approximation is iterative. It consists of the following steps at each iteration.

Step: 1. Use the $n + 1$ data points $t_1, t_2, \ldots, t_{n+1}$ determined from the previous iteration and set up the $n + 1$ linear simultaneous equations

$$f(t_i,\mathrm{x}) = F(t_i) - P(t_i,\mathrm{x}) = (-1)^i E \quad i = 1, 2, \ldots, n-1, \tag{4.45}$$

i.e.

$$\begin{bmatrix} F(t_1) \\ F(t_2) \\ \vdots \\ F(t_{n+1}) \end{bmatrix} - \begin{bmatrix} 1 & t_1 & t_1^2 & \ldots & t_{n-1} \\ 1 & t_2 & & & \vdots \\ \vdots & & & & \vdots \\ 1 & t_{n+1} & t_{n+1}^2 & \ldots & t_{n+1}^{n-1} \end{bmatrix} \begin{bmatrix} x_1 \\ x_2 \\ \vdots \\ x_n \end{bmatrix} = \begin{bmatrix} -E \\ E \\ \vdots \\ (-1)^{n+1}E \end{bmatrix} \tag{4.46}$$

Initially t_1 and t_{n+1} are taken to be the end points of the specified interval and the remaining $n - 1$ points are chosen within it.

2. Solve equation 4.46 for $x_1, x_2, \ldots, x_n$ and E. Theoretically, an exact solution is always possible. In practice, standard direct or iterative techniques for the numerical solution of linear equation systems must be applied, see for example Fox (1964).
3. Determine the values $t_2', t_3', \ldots, t_n'$ of t which correspond to the $n - 1$ extrema of the current error function given by $f(t, \mathrm{x})$ within the interval t_1 to t_{n+1}. The method used will depend on the form of F which may be either specified functionally or be given numerically. Often it will involve locating the zeros of the first derivative of the error function in the interval using a suitable non-linear optimization method such as Powell's least squares method given in the previous section. Alternatively, the error may simply be evaluated at a large number of points and the $n - 1$ values corresponding to separate peaks of the error function selected.
4. Choose a new set of data points t_2 to t_n to correspond to the error peaks determined in Step 3.

Retain the end points t_1 and t_{n+1}. (Although they may not be points of maximum error, they usually are.) Repeat the process iteratively by returning to Step 1 until E or the data points t_2 to t_n converge to within specified tolerances.

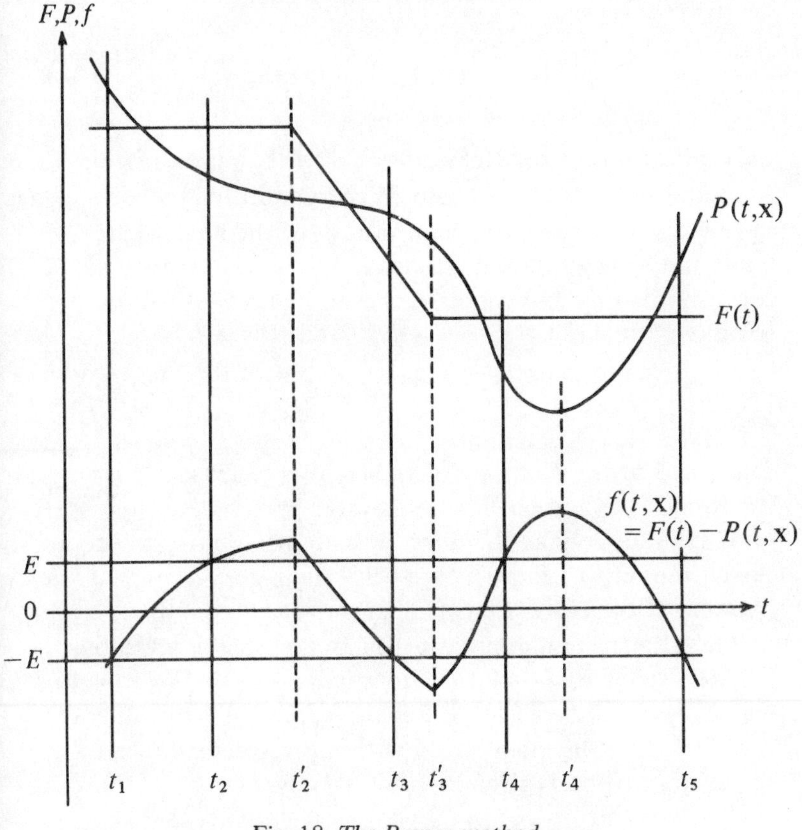

Fig. 18. *The Remez method.*

The process is illustrated in Figure 18. The effect of successive iterations is to progressively move the sample points to the position of the polynomial maxima and minima while setting the error magnitudes at these points all equal to $E_{\min}$. Note that at any stage $E_{\min}$ is sandwiched between the current value of E and the maximum value of the absolute error of the current error curve f. The sequence of error functions converges uniformly on the interval t_1 to t_{n+1}, but speed of convergence of the algorithm is dependent on the initial choice of $t_2, \ldots, t_n$ prior to the first iteration. The choice of equidistant points for example, is not recommended. Meinardus recommends the use of the extremal points of the Chebyshev polynomial $T_{n+1}(t)$ in the approximation interval $t_1 \leqslant t \leqslant t_{n+1}$, i.e. the points

$$t_i = \tfrac{1}{2}(t_{n+1} - t_1)\left(1 - \cos\frac{i\pi}{n+1}\right) + t_1, i = 0, 1, \ldots, n+1.$$

Applications of the Remez method with approximating functions other than polynomials is complicated by convergence problems. Nevertheless generalizations of the method are computationally fast and efficient and are widely used, see for example Fox and Parker, Chapter 4, and Meinardus, Chapters 7 to 11.

Zero shifting

The *zero shifting* method due to Maehly (1963) is similar to the Remez method but it successively approximates the minimax polynomial utilizing a polynomial approximation to the current error function to specify the zeros rather than the extrema of the current error function.

When the current approximating polynomial is given by $P(t, \mathrm{x})$ of equation 4.44, the equation corresponding to 4.46 which specifies the zeros of the corresponding error function at n data points $t_1, t_2, \ldots, t_n$ within the specified interval of approximation is given by

$$\begin{bmatrix} 1 & t_1 & t_1^2 \ldots t^{n-1} \\ \vdots & & \vdots \\ 1 & t_n & t_n^2 \ldots t_n^{n-1} \end{bmatrix} \begin{bmatrix} x_1 \\ \vdots \\ x_n \end{bmatrix} = \begin{bmatrix} F(t_1) \\ \vdots \\ F(t_n) \end{bmatrix}. \qquad (4.47)$$

This may be solved for $x_1, x_2, \ldots, x_n$ to obtain a polynomial which is an exact fit to F at $t_1, \ldots, t_n$. Suppose the resulting error function given by $f(t, \mathrm{x}) = F(t) - P(t, \mathrm{x})$ can be approximated by a polynomial of degree n. Then the error approximation can be represented in the form

$$e(t; t_1, \ldots, t_n) = \Pi_{i=1}^{n}(t - t_i). \tag{4.48}$$

Taking (natural) logarithms and the total derivative of e for fixed t yields

$$d \log |e| = \Sigma_{i-1}^{n} \frac{dt_i}{|t - t_i|}, \tag{4.49}$$

with finite approximation

$$\Delta \log |e| = -\Sigma_{i-1}^{n} \frac{\Delta t_i}{(t - t_i)}. \tag{4.50}$$

(The minus sign in 4.50 arises from the fact that the change in log $|e|$ will be taken to be a decrease.)

Denote the successive $n + 1$ maxima and minima of the current error function within the current interval of approximation including its two end points by $a_1, a_2, \ldots, a_{n+1}$. The steps for determination of minimum equal ripple error can then be formulated.

Step: 1. Solve equation 4.47 for $x_1, \ldots, x_n$ given the data points $t_1, \ldots, t_n$ within the specified interval. The latter are chosen arbitrarily for the first iteration.

2. Using a suitable non-linear technique determine the $n - 1$ maxima and minima of the current error function given by $f(t, \mathrm{x})$ within the approximation interval. Adding its end points yields $a_1, \ldots, a_{n+1}$. Also determine the error values $e_1, \ldots, e_{n+1}$ at these points.

3. An equal ripple amplitude E is required at each of the extrema of the error function. The required change in error at the points $a_1, \ldots, a_{n+1}$ can be determined from the equation

$$\Delta \log |e_j| = \log |E| - \log |e_j|, \tag{4.51}$$

making use of equation 4.50, as

$$-\sum_{i=1}^{n} \frac{\Delta t_i}{(a_j - t_i)} = \log|E| - \log|e_j|, \quad j = 1, \ldots, n+1 \tag{4.52}$$

These are $n + 1$ linear equations which must be solved for $\Delta t_1, \ldots, \Delta t_n$ and $\log |E|$.

4. Correct the data points $t_1, \ldots, t_n$ by adding the increments $\Delta t_1, \ldots, \Delta t_n$ found at Step 3 and update E. Return to Step 1 and iterate until convergence is obtained. The optimal polynomial approximation is found as $P(t, \mathbf{x}_{\min})$ by one more execution of Step 1 for $\mathbf{x}_{\min}$. The corresponding terminal error function will then have equal ripple of magnitude $E_{\min}$ found at the last execution of Step 3. Convergence properties are similar to those of the Remez method.

When convergent the two methods described above both converge to the minimax approximation to the specified function by a polynomial of degree $n - 1$. The minimal polynomial satisfies the equal ripple error criterion on the approximation interval. The methods are very similar and both require a first guess to a set of points in the approximation interval which is optimized by choice of the parameters using both linear and non-linear numerical techniques. In common with the general optimization methods with which this book is primarily concerned, more advanced techniques developed from the Remez and zero shifting methods for more general approximating functions often require a first approximation to the parameters involved, see for example Meinardus, Part II. Minimax methods contrast with (and often even involve) methods for the simpler least squares criterion. The two basic methods given here in a form suitable for polynomial approximation give some idea of the general complexity involved.

Exercises 4

1 Compare the performance of the advanced optimization methods in minimizing Rosenbrock's (1960) function

given by

$$E = 100(x_2 - x_1^2)^2 + (1 - x_1)^2 .$$

Start from (– 1.2, 1.0).
Use standard computer library subroutines.

2 Repeat with Powell's (1964) function given by

$$E = (x_1 + 10x_2)^2 + 5(x_3 - x_4)^2 + (x_2 - 2x_3)^4 + 10(x_1 - x_4)^4 .$$

Start from (3, – 1, 0, 1).

3 Repeat with Wood's function given by

$$E = 100(x_1^2 - x_2)^2 + (1 - x_1)^2 + 90(x_3^2 - x_4)^2 + (1 - x_3)^2$$
$$+ 10.1(x_2 - 1)^2 + 10.1(x_4 - 1)^2 + 19.8(x_2 - 1)(x_4 - 1).$$

Start from (–3, –1, –3, –1).

4 Repeat with Fletcher and Powell's (1963) helical valley given by

$$E = 100\,[(x_3 - 10\theta)^2 + (r - 1)^2] + x_3^2$$

where

$$2\pi\theta = \tan^{-1}(x_2/x_1)$$
$$r = (x_1^2 + x_2^2)^{1/2}$$

Start from (–1, 0, 0).

5 Repeat with the function

$$E = [10 - (x_1 - 1)^2 - (x_2 - 1)^2]^2 + (x_1 - 2)^2 + (x_2 - 4)^2$$

from the exercises to Chapter 3.
Start from (1, 0).

6 Repeat with Parkinson and Hutchinson's (1972) functions given by

$$E = \Sigma_{i=1}^{n} i\, x_i^2 + n x_n^4$$

and

$$E = n^{-1} \Sigma_{i=1}^{n} (\cosh x_i + x_i^4)$$

for a range of values of n. Compare the performance of the algorithms on the two functions as n increases.

7 Approximate the function given by

$$y = 2t \quad \text{for} \quad t \leqslant \tfrac{1}{2}$$

$$y = 2(1-t) \quad \text{for} \quad t > \tfrac{1}{2}$$

in the interval $0 \leqslant t \leqslant 1$. Compare results from the Remez and Zero Shifting methods.

CHAPTER FIVE

Constrained optimization

5.1 Introduction

Constrained optimization was introduced in Section 1.2 where it was sometimes found necessary for practical reasons arising in applications to search for the optimum of a function f within a certain *feasible region* defined by inequality and/or equality constraints of the form

$$g_i(\mathbf{x}) \leqslant 0 \qquad i = 1, 2, \ldots, l \tag{5.1}$$

$$h_j(\mathbf{x}) = 0 \qquad j = 1, 2, \ldots, m. \tag{5.2}$$

Several basic approaches have been used to handle constraints using the unconstrained optimization techniques so far considered. Some of these will be described below, but since the subject of constrained optimization is a large and growing one, the treatment given here can serve only to introduce the topic and point out some of the difficulties.

For problems with linear, quadratic, or convex objective functions and linear constraints there exist efficient procedures which make use of the special structure of the problem. Similar remarks apply to the situation in which all the functions in the problem are (or can be transformed to be) convex, the sum of convex functions of a single variable, or the sum of polynomial expressions in several variables. The techniques for such problems constitute the bulk of the computational aspects of constrained optimization, or *mathematical programming*. They are discussed adequately in other texts, see for example Wilde and Beightler (1967) and Luenberger (1972). Accordingly we shall concentrate attention here on algorithms for problems whose objective and constraints involve general non-linear functions.

The most obvious method of dealing with a constraint is to convert a constrained problem into a new problem which is not constrained. This can sometimes be done in a way in which the original constraints cannot be violated in the transformed problem. A simple example of this, involving inequality constraints, concerns the minimization of the objective function given by $E = f(\mathbf{x})$ for positive values of $\mathbf{x}$. A new set of variables defined by $x_i = y_i^2$ converts this problem into an unconstrained minimization in terms of $\mathbf{y}$. A more complicated transform, namely $2x_i = (x_{U,i} + x_{L,i}) + (x_{U,i} - x_{L,i}) \sin y_i$, is needed to remove a constraint of the form $x_{L,i} \leqslant x_i \leqslant x_{U,i}$. A third example concerns the minimization of f subject to linear equality constraints of the form

$$\mathbf{K}_1\mathbf{y} + \mathbf{K}_2\mathbf{z} = \mathbf{b}, \tag{5.3}$$

where $\mathbf{K}_2$ is a square non-singular matrix. Then solving for $\mathbf{z}$ in terms of $\mathbf{y}$ converts the problem to unconstrained minimization in $\mathbf{y}$ of the objective function given by

$$E = f(\mathbf{y}, \mathbf{K}_2^{-1}\mathbf{b} - \mathbf{K}_2^{-1}\mathbf{K}_1\mathbf{y}) \tag{5.4}$$

We shall study a generalization of this method in discussing the 'reduced gradient' method. For the moment note that these examples emphasize that the problem should be properly formulated, and that the opportunity may be taken to remove constraints when this is possible. On the other hand, new local minima are often introduced in situations like the second example.

5.2 The Kuhn–Tucker conditions

The classical method of removing equation constraints of the form 5.2 is due to Lagrange. He considered the function given by

$$f(\mathbf{x}) + \Sigma_{j=1}^{m} v_j h_j(\mathbf{x}), \qquad (5.5)$$

involving the 'undetermined' or *Lagrange multipliers* v_j, $j = 1, \ldots, m$; requiring it to be stationary in the $n + m$ variables $x_1, \ldots, x_n; v_1, \ldots, v_m$. When all the problem functions are continuously differentiable it may be shown, under suitable further conditions on the constraint functions $h_1, \ldots, h_m$, that a necessary condition for $\bar{\mathbf{x}}$ to be a local constrained minimum of f is that there exist multipliers $\bar{v}_1, \ldots, \bar{v}_m$ such that

$$\nabla f(\bar{\mathbf{x}}) + \Sigma_{i=1}^{m} \bar{v}_j \nabla h_j(\bar{\mathbf{x}}) = \mathbf{0}, \qquad (5.6)$$

see, for example, Apostal (1957), Chapter 7. Note that the stationarity conditions involving the partial derivatives of 5.5 with respect to the v_j's yield the constraints 5.2. The classical approach to constrained optimization is therefore to apply an appropriate optimization technique to solve the system of $m + n$ non-linear equations given by 5.6 and 5.2. Often this will be accomplished by applying a least squares routine to the sum of squares of the left hand sides of 5.6 and the constraints 5.2. The inequality constraints 5.1 can be handled by introducing *slack variables* y_i to yield the constraints $g_i(\mathbf{x}) + y_i^2 = 0, i = 1, \ldots, l$. Aside from the problem of distinguishing local constrained minima from other constrained optima, this method is usually computationally highly inefficient when applied to many constraints.

The result concerning the stationarity of the *Lagrangian* function 5.5 was first extended to inequality constraints by Kuhn and Tucker (1951). As in the classical case, some conditions on the functions of the problem are required. For simplicity we shall from this point assume that the following *regularity condition* is satisfied at each constrained local minimum $\mathbf{x}_{\min}$.

Let I denote the set of indices i for which $g_i(\mathbf{x}_{\min}) = 0$ and suppose all the problem functions are differentiable. Then the row vectors $\nabla g_i(\mathbf{x}_{\min})$, for i in I, and $\nabla h_j(\mathbf{x}_{\min})$, $j = 1, \ldots, m$ are linearly independent.

A problem satisfying this assumption is termed (*locally*) *regular* and the constraints indexed by I are called the set of constraints *active* at $\mathbf{x}_{\min}$.

Under the regularity condition a necessary condition that $\bar{\mathbf{x}}$ be a constrained local minimum of f is that there exist multipliers $\bar{u}_1, \ldots, \bar{u}_l; \bar{v}_1, \ldots, \bar{v}_m$ such that

$$\nabla f(\mathbf{x}) + \Sigma_{i=1}^{l} \bar{u}_i \nabla g_i(\bar{\mathbf{x}}) + \Sigma_{j=1}^{m} \bar{v}_j \nabla h_j(\bar{x}) = 0 \qquad (5.7)$$

$$\bar{u}_i g_i(\bar{\mathbf{x}}) = 0 \qquad i = 1, \ldots, l \qquad (5.8)$$

$$\bar{u}_i \geqslant 0 \qquad i = 1, \ldots, l. \qquad (5.9)$$

For a proof and thorough discussion of the relation of regularity conditions to this result see Dempster (1975), Chapter 7. See also Luenberger (1972), Chapter 10. Conditions 5.7 to 5.9 are known as the *Kuhn–Tucker optimality* conditions. Together with 5.2 they are the first order necessary conditions characterizing a local constrained minimum. The relations 5.8 are termed *complementary slackness* conditions since together with 5.9 they imply that at an optimum a positive multiplier corresponds to an inequality constraint *binding*, i.e. holding as an equation, while a non-binding constraint corresponds to a zero multiplier.

The Lagrange multipliers $u_1, \ldots, u_l; v_1, \ldots, v_m$ can usually be given a practical interpretation related to the variables of the problem. In physical problems the multipliers may be momenta, power, etc., while in economic or management problems they are usually prices. They are often referred to as *dual variables* due to the fact that in many special cases, for example when all the problem functions are linear or convex, they form some or all of the variables of a maximization problem *dual* to the given minimization problem in the sense that the objective function values of the two problems agree and the complementary slackness relations hold between their variables. For a discussion of the extensive duality theory concerning such *mathematical programming* problems the reader is referred to Wilde and Beightler (1968) or Dempster. The dual programme to a given problem can often be interpreted to express an optimization in terms of the dual variables which is implicit in the original problem. Duality principles such as the minimum energy principle have a long history in mathematical physics. For these reasons it is an important asset of a numerical procedure for constrained optimization that it leads easily to a calculation of the optimal Lagrange multipliers.

The Kuhn–Tucker conditions 5.7 to 5.9 may be regarded as a modification of the vanishing of the gradient of f at a local unconstrained minimum to take account of local constrained optima on the constraint boundaries. They express the fact that for the optimum values $\bar{\mathbf{u}}$, $\bar{\mathbf{v}}$ of the Lagrange multipliers, the Langrangian given by

$$\phi(\mathbf{x}; \mathbf{u}, \mathbf{v}) = f(\mathbf{x}) + \Sigma_{i=1}^{l} u_i g_i(\mathbf{x}) + \Sigma_{j=1}^{m} v_j h_j(\mathbf{x}), \quad (5.10)$$

must have a local stationary point which corresponds to the given constrained minimum $\bar{\mathbf{x}}$. Similarly, the second order conditions for a local unconstrained minimum can be generalized to the constrained situation. This is effected by a suitable modification of the argument based on the second term of the Taylor series expansion about the minimum, applied to the Lagrangian. If the problem functions are twice continuously differentiable, it can be shown that at a local constrained minimum $\bar{\mathbf{x}}$ the inequality

$$\mathbf{x}^{\mathrm{T}} [\nabla^2 f(\bar{\mathbf{x}}) + \Sigma_{i=1}^{l} \bar{u}_i \nabla^2 g_i(\bar{\mathbf{x}}) + \Sigma_{j=1}^{m} v_j \nabla^2 h_j(\bar{\mathbf{x}})] \, \mathbf{x} \geqslant 0, \quad (5.11)$$

involving the Hessian of the Lagrangian, must hold for any $\bar{\mathbf{u}}$ and $\bar{\mathbf{v}}$ satisfying the Kuhn–Tucker conditions and any $\mathbf{x}$ such that

$$\nabla g_i(\bar{\mathbf{x}}) \, \mathbf{x} = 0 \qquad i \in I, \text{ i.e. } g_i(\bar{\mathbf{x}}) = 0 \quad (5.12)$$

$$\nabla h_j(\bar{\mathbf{x}}) \, \mathbf{x} = 0 \qquad j = 1, \ldots, m. \quad (5.13)$$

Equations 5.12 and 5.13 restrict $\mathbf{x}$ to lie in the intersection of the hyperplanes tangent to the constraints satisfied as equations at $\bar{\mathbf{x}}$. A sufficient condition that $\bar{\mathbf{x}}$ be locally unique, i.e. an *isolated* local constrained minimum, is that for any $\mathbf{x}$ such that

$$\nabla g_i(\bar{\mathbf{x}})\,\mathbf{x} = 0 \qquad \text{when } \bar{u}_i > 0 \tag{5.14}$$

$$\nabla g_i(\bar{\mathbf{x}})\,\mathbf{x} \geqslant 0 \qquad \text{when } u_i = 0, \text{ but } g_i(\bar{\mathbf{x}}) = 0 \tag{5.15}$$

$$\nabla h_j(\bar{\mathbf{x}})\,\mathbf{x} = 0 \qquad j = 1, \ldots, m, \tag{5.16}$$

the strict inequality

$$\mathbf{x}^{\mathrm{T}}[\nabla^2 f(\bar{\mathbf{x}}) + \Sigma_{i=1}^{l}\,\bar{u}_i\nabla^2 g_i(\bar{\mathbf{x}}) + \Sigma_{j=1}^{m}\,\bar{v}_j\nabla^2 h_j(\bar{\mathbf{x}})]\,\mathbf{x} > 0 \tag{5.17}$$

must hold. For proofs of these relations see Fiacco and McCormick (1968), Chapter 2, or Luenberger (1972), Chapter 10.

The convergence properties of constrained optimization techniques can be discussed in terms of whether they guarantee convergence to a point satisfying the Kuhn–Tucker first order conditions 5.7 to 5.9 or to one satisfying the second order conditions 5.11 to 5.13 as well. The terms *first order* and *second order* method are used to distinguish the two properties. If the rate of convergence of a second order method is to be acceptable it will frequently be designed to converge to an isolated local constrained minimum satisfying the slightly stronger second order conditions 5.14 to 5.17. This is analogous to designing the approximation to the Hessian matrix to be positive definite in a quasi-Newton method for unconstrained optimization, cf. Section 4.4.

When all the problem functions are convex as well as differentiable, the first order conditions 5.7 to 5.9 are sufficient to ensure that a point $\bar{\mathbf{x}}$ satisfying them is a global constrained minimum. In that situation such a minimum corresponds, for a regular problem, to a saddle-point of the Lagrangian ϕ (given by 5.10) which is convex in $\mathbf{x}$ and linear in the dual variables $\mathbf{u}$, $\mathbf{v}$. The stationarity of the Lagrangian at $\bar{\mathbf{x}}$ and $\bar{\mathbf{u}}$, $\bar{\mathbf{v}}$ is therefore sufficient to ensure that $\bar{\mathbf{x}}$ minimizes ϕ in $\mathbf{x}$ at $\mathbf{u} = \bar{\mathbf{u}}$, $\mathbf{v} = \bar{\mathbf{v}}$, while $\bar{\mathbf{u}}$, $\bar{\mathbf{v}}$ maximizes ϕ in $\mathbf{u}$, $\mathbf{v}$ at $\mathbf{x} = \bar{\mathbf{x}}$, as is required at a saddle point, see Section 1.4.

5.3 Constrained optimization techniques

Methods for incorporation of constraints by modification of unconstrained techniques fall into two groups.

(i) *Methods which take explicit account of the constraints at each step.*

These are usually of two types. *Direct search* methods recognize when the current evaluation point is near a constraint and modify the directions of search, possibly by means of penalty function corrections to the objective function, accordingly. *Small step gradient* methods modify the method of steepest descent to deflect the direction given by the negative gradient into the feasible region when a constraint is violated. In these methods constraints are treated as reflecting barriers. An advantage of search methods over gradient methods is that objective function gradients in the constraint region can sometimes prove unreliable. We shall describe modifications of pattern search and the method of rotating co-ordinates before discussing small step gradient methods briefly.

(ii) *Methods which take implicit account of the constraints using global penalty functions.*

The idea is to penalize constraint violation by adding a sequence of penalty functions to the objective function in such a way that the solutions to the resulting sequence of unconstrained problems tend to a constrained minimum. This approach results in a class of techniques termed *sequential unconstrained* methods. Any unconstrained technique may be used to solve the unconstrained problems.

Both types of method have disadvantages and the choice of method is problem dependent. Methods in the first group become inefficient when constraints are complicated since the search must then be continually modified to find its way around the constraint surfaces. Methods in the second group attempt to minimize a function other than the objective into which all kinds of peculiarities may have been introduced artificially. The functions created are often those most difficult to optimize. Methods in Group II are however the most successful for general applications, while Group I methods are more suitable for simple situations.

Modification of unconstrained methods is not the only approach to non-linear constrained optimization and several techniques available specifically consider the constrained problem. The most effective and widely used of these are known as *large step gradient methods* since they modify the direction of steepest descent to take explicit account of constraints. Like the methods for minimax approximation discussed in the previous chapter, they make use of an unconstrained minimization technique at each iteration. We shall outline two principal variants of this approach, the *gradient projection* and the *reduced gradient* methods.

Recently a number of promising methods have been proposed which attempt to solve the constrained problem by minimizing a version of the Lagrangian function of the problem in either a single unconstrained optimization or a sequence of problems subject to linear constraints. These *Lagrangian* methods may be divided into two classes: those making use of approximations to the problem functions and those using approximations to the Lagrange multipliers to obtain a global penalty function. We shall describe both approaches briefly.

5.4 Direct search methods with constraints

Multiple gradient summation

The pattern search is one of the more successful basic minimization techniques and a number of extensions have been suggested for the incorporation of constraints. The method briefly described here is a simple modification of the basic method which modifies the search direction and retains the original function. It is of practical use only for problems with few inequality constraints. The merit of the method lies in its ease of application and its consequent suitability for exercise or demonstration.

The unconstrained pattern search, described fully in Section 3.2, consists of a sequence of local explorations about a base to find a suitable direction of movement, followed by pattern moves in that direction to establish a new base. When a constraint interrupts a pattern move the conventional pattern search must cease on the constraint boundary. One method of continued search due to Klingman and Himmelblau (1964), called the *multiple gradient summation* technique, is to make all future movements on the assumption that the minimum lies near to or on the constraint boundary.

When constraints are encountered, the multiple gradient summation technique defines a new direction of search which is a linear combination of the function gradient and the violated constraint gradients, viz.

$$-\frac{\nabla \mathbf{f}^{\mathrm{T}}}{|\nabla \mathbf{f}|} - \sum_{i=1}^{P} \frac{\nabla \mathbf{g}_i^{\mathrm{T}}}{|\nabla \mathbf{g}_i|}, \tag{5.18}$$

where the sum is taken over the violated constraints evaluated at the current point. When this technique is used with pattern search the function gradient $\mathbf{g} = \nabla \mathbf{f}^{\mathrm{T}}$ is approximated by the direction vector of the predicted pattern move. The constraint function gradient vectors $\nabla \mathbf{g}_i^{\mathrm{T}}$ are usually evaluated analytically.

Rotating co-ordinates

The method of rotating co-ordinates for unconstrained minimization, described in Section 4.3, has been developed further by Rosenbrock (1960) to include constraints of the form

$$x_{\mathrm{L},i} \leqslant x_i \leqslant x_{\mathrm{U},i} \qquad i = 1, 2, \ldots, l, \;\; l \geqslant n, \tag{5.19}$$

where $x_{n+1}, x_{n+2}, \ldots, x_l$ are functions of $\mathbf{x}$ and $x_{\mathrm{L},i}$ and $x_{\mathrm{U},i}$ *may* be functions of $\mathbf{x}$. The first n constraints therefore limit the values of the parameters $x_1, x_2, \ldots, x_n$ and all are assumed present. This involves no essential loss of generality for inequality constraints since the functional form of the limits includes constant limits which can usually be inserted without difficulty even when no limit is specified by the actual problem. The last $l-n$ constraints are of the form $x_{\mathrm{L},i} \leqslant f_i(x) \leqslant x_{\mathrm{U},i}$, $i = n+1, n+2, \ldots, l$. These may not be present, but when they are they must be few in number for the method to be efficient.

A boundary region is associated with each constraint, with a width δ_i defined by an expression such as $10^{-4}(x_{U,i} - x_{L,i})$. The variable x_i is then within the constraint region if either

$$x_{L,i} \leqslant x_i \leqslant x_{L,i} + \delta_i, \tag{5.20}$$

or

$$x_{U,i} \geqslant x_i \geqslant x_{U,i} - \delta_i. \tag{5.21}$$

A depth of penetration λ is defined for the lower boundary as

$$\lambda = \frac{(x_{L,i} + \delta_i) - x_i}{\delta_i} \tag{5.22}$$

and for the upper boundary as

$$\lambda = \frac{x_i - (x_{U,i} - \delta_i)}{\delta_i} \tag{5.23}$$

depending on which boundary region is entered.

The procedure for using rotating co-ordinates is only slightly modified to deal with constraints in the form 5.19. The search proceeds normally from a first approximation in the feasible region, but not within the boundary zones. If the search crosses the boundary into the infeasible region the test in the current direction is assumed to fail. If the search enters a boundary zone the computed value E of the objective function f is replaced by a new value given by

$$E' = E + (E^* - E)(1 - 3\lambda + 4\lambda^2 - 2\lambda^3), \tag{5.24}$$

where E^* is the least value of E obtained before a boundary region was entered. If further boundary regions are entered then E' is similarly modified further. Except for the treatment of infeasible test points and the modifications to the values of the objective function the search proceeds as for the unconstrained case.

Rosenbrock's method for constrained minimization is a local penalty function technique since the objective function is altered in the vicinity of a constraint to create a local minimum of the function within the boundary region. The effect of this is to construct an artificial valley close to binding constraints which the minimization process will follow. The same result is effected directly by deflection of the next evaluation point in small step gradient methods to which we shall next turn briefly. The use of the boundary region for search is not applicable to equality constraints or to other methods of minimization and a more general method of altering the objective function is required. However, although not efficient for general problems, the modifications described operate successfully with search by rotating co-ordinates and are easy to incorporate into the standard computations.

5.5 Small step gradient methods

In Section 3.3 the basic iteration of the method of steepest descent was given as

$$\mathbf{x}_{k+1} = \mathbf{x}_k + \lambda\mathbf{u}$$

where $\mathbf{u} = -\mathbf{g}/|\mathbf{g}|$, $\mathbf{g}^{\mathrm{T}} = \nabla f(\mathbf{x}_k)$ and the value of λ is to be determined by a linear search. The direction $\mathbf{u}$ can be modified in the presence of the inequality constraints 5.1 by using penalty functions for constraint violation as follows. (Without loss of generality we neglect normalization of the search direction to simplify the formulae). Write

$$\mathbf{u}^{\mathrm{T}} = -\nabla f(\mathbf{x}_k) - \Sigma_{i=1}^{l} w_i(\mathbf{x}_k)\nabla g_i(\mathbf{x}_k), \qquad (5.25)$$

where $w_i(\mathbf{x}_k) = W$, a suitably large positive constant, if $g_i(\mathbf{x}_k) > 0$, and $w_i(\mathbf{x}_k) = 0$ if $g_i(\mathbf{x}_k) \leqslant 0$. Note the similarity with the multiple gradient summation technique for pattern search, cf. 5.18. Since $\nabla g_i(\mathbf{x}_k)$ is the outward pointing normal to the constraint surface given by $g_i(\mathbf{x}) = 0$, the term $w_i(\mathbf{x}_k)\nabla g_i(\mathbf{x}_k)$ deflects the linear search at the kth iteration back into the constraint set. Otherwise it would tend to leave under the influence of $-\nabla f(\mathbf{x}_k)$.

Let us suppose that this modified first order process for the inequality constrained problem terminates successfully at a value $\bar{\mathbf{x}}$ after K iterations. Assume that during the last J iterations only small changes in $\mathbf{x}$, E and λ are made. Then letting $I = K - J$,

$$\begin{aligned}\bar{\mathbf{x}} &= \mathbf{x}_I + \Sigma_{k=I+1}^{K} \lambda_k [-\nabla f(\mathbf{x}_k) - \Sigma_{i=1}^{l} w_i(\mathbf{x}_k) \nabla g_i(\mathbf{x}_k)] \\ &\approx \mathbf{x}_I - \lambda_K [J f(\bar{\mathbf{x}}) + \Sigma_{i=1}^{l} \Sigma_{k=I+1}^{K} w_i(\mathbf{x}_k) \nabla g_i(\bar{\mathbf{x}})]. \quad (5.26)\end{aligned}$$

If $g_i(\bar{\mathbf{x}}) \ll 0$, then clearly $w_i(\mathbf{x}_k) \approx w_i(\bar{\mathbf{x}}) = 0$ over the last j iterations; while if $g_i(\bar{\mathbf{x}}) \approx 0$, $w_i(\mathbf{x}_k)$ will have taken the value W over a number of them. Hence it follows from 5.26, rearranging and dividing by J, that

$$\mathbf{0} \approx \frac{1}{J}(\bar{\mathbf{x}} - \mathbf{x}_I) = -\lambda_K \left[\nabla f(\bar{\mathbf{x}}) + \Sigma_{i=1}^{l} \frac{1}{J} \Sigma_{k=I+1}^{K} w_i(\mathbf{x}_k) \nabla g_i(\bar{\mathbf{x}})\right]$$

or

$$\nabla f(\bar{\mathbf{x}}) + \Sigma_{i=1}^{l} \bar{u}_i \nabla g_i(\bar{\mathbf{x}}) \approx \mathbf{0}. \quad (5.27)$$

where

$$\bar{u}_i = \frac{1}{J} \Sigma_{k=I+1}^{K} w_i(\mathbf{x}_k) \geqslant 0 \quad (5.28)$$

and

$$\bar{u}_i g_i(\bar{\mathbf{x}}) \approx 0. \quad (5.29)$$

But 5.27 to 5.29 are approximately the version of the Kuhn–Tucker conditions 5.7 to 5.9 appropriate to the inequality constrained problem. Hence the optimal multipliers are approximated by 5.28.

Although this modification of steepest descent is inappropriate to equation constraints, methods based on linearization of equation constraints and a similar result concerning optimal dual variables are available. In common with their unconstrained counterparts, however, first order gradient methods have proved relatively inefficient due to slow convergence. We shall not discuss them further, but the interested reader is referred to Greenstadt (1966) and to the survey of Wolfe (1967).

5.6 Sequential unconstrained methods

A sequential unconstrained method for minimizing f subject to inequality and equality constraints 5.1 and 5.2 associates with the problem a sequence of unconstrained minimizations of functions given by $F^k(\mathbf{x})$, $k = 1, 2, \ldots$, which are defined in terms of the problem functions f, $\mathbf{g}$ and $\mathbf{h}$. The functions F^k are constructed so that the sequence of solutions $\mathbf{x}_k$ to the unconstrained problems tend to a constrained minimum of f.

Penalty functions

The most successful of these sequential techniques involve penalty functions for constraint violation. (See Fiacco and McCormick (1968) or the survey of Lootsma (1972a) for discussions of the alternative 'method of centres'.) These methods define $F^k(\mathbf{x})$ as

$$f(\mathbf{x}) + \Sigma_{i=1}^{l} w_i G(g_i(\mathbf{x}), r) + \Sigma_{j=1}^{m} H(h_j(\mathbf{x}), r), \qquad (5.30)$$

where F^k depends on k through the parameters $w_1, \ldots, w_l$ and r which may be changed for each minimization in the sequence. Most methods so far proposed set w_i equal to a positive constant when the ith inequality constraint boundary is approached and to zero otherwise. Usually a sequence of r values is taken which decreases to zero as k tends to infinity.

The inequality constrained modification of the method of steepest descent treats a special case of 5.30 in which $w_i = w_i(\mathbf{x}_k)$, $G(g_i(\mathbf{x}), r) = g_i(\mathbf{x})$ and H is the zero function. The procedure makes a single linear search in the direction $-\nabla F^k$ to estimate $\mathbf{x}_k$.

The earliest proposals for G and H involved quadratic functions of the form

$$G(g_i(\mathbf{x}), r) = r^{-1}[-g_i(\mathbf{x})]^2 \qquad (5.31)$$

and

$$H(h_j(\mathbf{x}), r) = r^{-1} h_j^2(\mathbf{x}). \qquad (5.32)$$

(The minus sign occurs in 5.31 to take account of the direction of the inequality in the constraint $g_i(\mathbf{x}) \leqslant 0$.)
Any non-zero weights w_i are taken equal to unity. Methods using functions, such as 5.31, which only penalize constraint violations are termed *exterior* penalty function methods. With these methods the successive unconstrained minima $\mathbf{x}_k$ lie outside the feasible region until r_k^{-1} tending to infinity

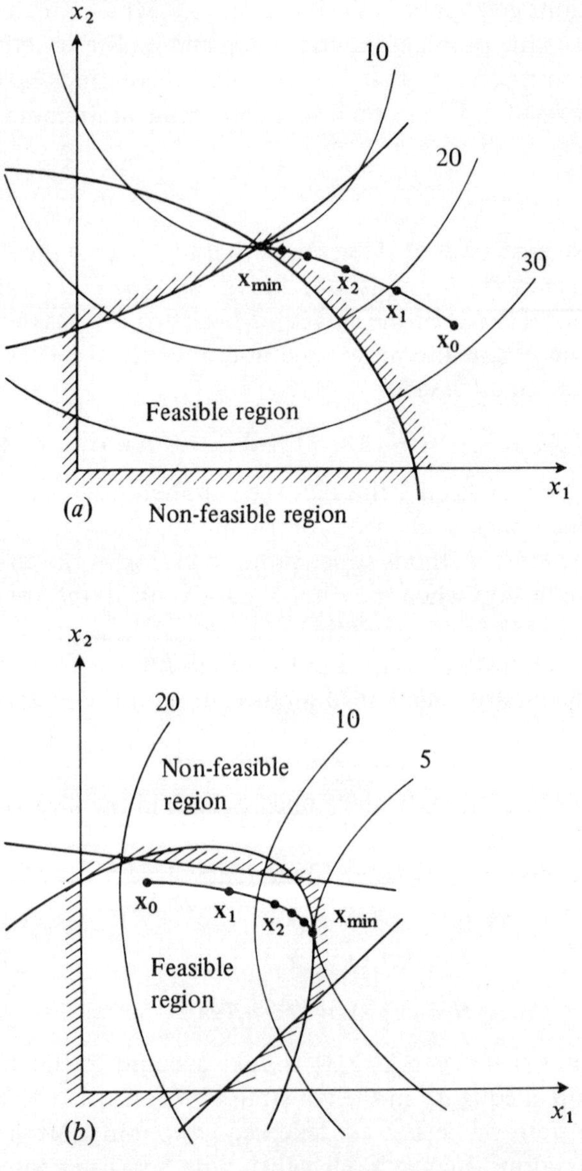

Fig. 19. *Trajectories of sequential unconstrained minimization techniques. Diagram (a). Exterior point penalty function. Diagram (b). Interior point penalty function.* ///////// *is the constraint boundary.*

forces a constrained minimum to be approximated. A typical trajectory is illustrated in Diagram A of Figure 19. *Exterior point* penalty functions, or *loss* functions, are clearly necessitated by equation constraints. Loss functions H of quadratic type have proved effective in this situation and their use is backed by extensive theoretical investigation, see for example Fiacco and McCormick.

When inequality constraints are present, *interior point* penalty functions, or *barrier* functions, are computationally preferable. For fixed r, such G functions are non-zero in the interior of the feasible region (defined by the inequalities $g_i(\mathbf{x}) \leqslant 0, i = 1, \ldots, l$) and infinite on its boundaries. They maintain successive $\mathbf{x}_k$ in the interior of the feasible region defined by the inequality constraints (see Diagram B of Figure 19) but require an initial approximation within that region, which must be a connected set. As r tends to 0 the constrained minimum is approximated. The most popular barrier functions G are given by

$$-r \ln [-g_i(\mathbf{x})], \tag{5.33}$$

$$r/[-g_i(\mathbf{x})] \tag{5.34}$$

and

$$r/[-g_i(\mathbf{x})]^2. \tag{5.35}$$

Note that each function is the partial derivative of the next with respect to g_i except for the appropriate sign correction. As depicted in Figure 20, successively faster approaches to infinity are made by 5.33 to 5.35 at the constraint boundaries where $g_i(\mathbf{x}) = 0$. In fact Lootsma (1972a) has suggested a classification of barrier functions G which possess a continuous second partial derivative with respect to g_i in terms of the order of the pole, i.e. infinite singularity, of $\partial G/\partial g_i$ at 0. Thus the barrier functions given by 5.33, 5.34 and 5.35 are of order 1, 2 and 3 respectively. (A similar classification of loss functions may be given in terms of the order of the zero of $\partial H/\partial g_i$ at 0; for example, the loss function given by 5.32 is of order 1.)

A class of barrier functions based on iterated logarithms and translates of

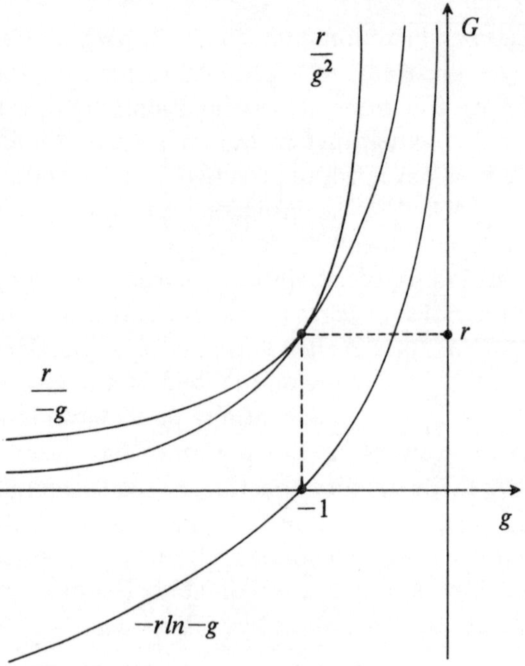

Fig. 20. *Interior point penalty functions.*

$$\ln\{K_i - \ln[-g_i(x)]\},$$

where K_i is a suitable constant, has been proposed by Osborne and Ryan (1970), and Zangwill (1967) studies the piece-wise linear loss function based on

$$\min\{0, -g_i(x)\}$$

However, in the first case, the poles, and in the second, the zero of these functions are not of finite positive order. The resulting disadvantage is that the convergence accelerating techniques described briefly below cannot be used with them.

The barrier function given by 5.34 was proposed by Carrol (1961). He termed a *created response surface* the surface generated by the corresponding F^k functions given by

$$f(\mathbf{x}) + r_k \sum_{i=1}^{l} \frac{w_i^k}{g_i(\mathbf{x})} \tag{5.36}$$

The resulting technique has come to be known as the *created response surface technique.*

Fiacco and McCormick favour 5.33 for ease in computing the inverse Hessian matrix when second order algorithms are used to minimize F^k. (For a similar reason they favour the exterior point G and H given by 5.31 and 5.32 for use with, respectively, linear equality and equation constraints.) Box, Swann and Davies (1969) have reported improved convergence to *global* minima using 5.35.

When both inequality and equation constraints are present in the problem, a *mixed* interior-exterior point penalty function made up of the sum of a barrier and a loss function is recommended. Fiacco and McCormick discuss a sum of 5.32 and 5.33 with any non-zero weights unity, but there is empirical evidence that the mixed penalty term

$$r^3 \sum_{i=1}^{l} \frac{w_i}{g_i^2(\mathbf{x})} + r^{-1}\sum_{j=1}^{m} h_j^2(\mathbf{x}) \tag{5.37}$$

is superior in spite of the increase in arithmetic operations needed to compute partial derivatives of F^k. A power r^2 multiplying the reciprocal terms is recommended for use with 5.34. In general there is both theoretical and demonstrated empirical advantage to be gained by using an r^p factor in barrier functions of order p and an r^{-q} factor in loss functions of order q according to Lootsma's classification. Box *et al* indicate faster convergence when all problem functions are scaled to the same order of magnitude. This suggests that unequal non-zero weights w_i would yield better results, but no really effective scheme for the automatic generation of the w_i so far exists.

Theoretical convergence to a local constrained minimum $\bar{\mathbf{x}}$ as r tends to zero has been established for the sequential unconstrained procedure using any of the penalty functions above. The problem functions are required to be continuous and the set of constrained local minima must be closed and bounded, see Fiacco and McCormick, Chapters 3 and 4 or Lootsma (1972a). The penalty terms tend to zero in the limit. Convergence under weaker conditions on the problem functions has been established for quadratic penalty functions by Beltrami (1969), (1970), Chapter 2. When r^p enters a barrier function of order p it can be shown that convergence of the unconstrained minima $\mathbf{x}_k$ (of F^k) to $\bar{\mathbf{x}}$ is of the order of the convergence of the r_k sequence to zero (see, for example, Osborne and Ryan, 1970).

Suppose the problem functions twice continuously differentiable and consider the gradient of F^k given by

$$\nabla F^k(\mathbf{x}) =$$
$$f(\mathbf{x}) + \Sigma_{i=1}^{l} w_i^k G_1(g_i(\mathbf{x}), r_k)\nabla g_i(\mathbf{x}) + \Sigma_{j=1}^{m} H_1(h_j(\mathbf{x}), r)\nabla h_j(\mathbf{x}). \tag{5.38}$$

Here G_1 and H_1 denote the partial derivatives of G and H with respect to their first arguments. Recall that $\mathbf{x}_k$ denotes the unconstrained minimum of F^k and define

$$u_i^k = w_i^k G_1(g_i(\mathbf{x}_k), r_k) \geqslant 0 \qquad i = 1, \ldots, l \tag{5.39}$$

and

$$v_j^k = H_1(h_j(\mathbf{x}_k), r_k) \qquad j = 1, \ldots, m. \tag{5.40}$$

It follows that

$$\nabla F^k(\mathbf{x}_k) = f(\mathbf{x}_k) + \Sigma_{i=1}^{l} u_i^k \nabla g_i(\mathbf{x}_k) + \Sigma_{j=1}^{m} v_j^k \nabla h_j(\mathbf{x}_k) \tag{5.41}$$
$$= \mathbf{0}$$

so that in the limit as k tends to infinity $\lim_{k\to\infty} \nabla F^k(\mathbf{x}_k) = 0$ and and the Kuhn–Tucker condition 5.7 is satisfied by any limiting $\bar{\mathbf{x}}$, $\bar{\mathbf{u}}$ and $\bar{\mathbf{v}}$. The other conditions 5.8 and 5.9 are automatically satisfied in virtue of 5.39 and the nature of the w_i. Similarly, the non-negative definiteness of the Hessian matrix of F^k at the unconstrained minimum $\mathbf{x}_k$ can be shown to lead to limiting $\bar{\mathbf{x}}$, $\bar{\mathbf{u}}$ and $\bar{\mathbf{v}}$ satisfying the second order conditions 5.11 to 5.13, see Fiacco and McCormick, Chapter 3. It can be shown (see Beltrami (1969) and Osborne and Ryan, 1970) that *irregular* situations in which the Kuhn–Tucker conditions do not hold at the constrained optimum are indicated by the convergence of all the u_i^k and v_j^k sequences to infinity.

A mixed penalty function algorithm

A sequential unconstrained minimization technique using a mixed penalty function consists of the following steps.

Step: 1. Find an initial approximation $\mathbf{x}_0$ in the interior of the feasible region for the inequality constraints, i.e. such that $g_i(\mathbf{x}_0) < 0, \ i = 1, \ldots, l$.
If no $\mathbf{x}_0$ is known, a few iterations (i.e. linear searches) of the modified gradient method (of the previous section) with large enough W will usually suffice to generate a suitable point. Fiacco and McCormick (1968) outline a method using more complicated penalty functions.

2. Determine r_1, the initial value of r.
This must be set so that the contribution to F^1 from the penalty term does not swamp the contribution due to f, forcing $\mathbf{x}_1$ well away from the inequality constraint boundary. On the other hand, r_1 must not be set so small that before the constrained minimum has been approximately located the constraints have little effect. When r is small, convergence to the minimum of F^k can be very slow. Fiacco and McCormick recommend a linear search over values of r_1 to minimize $|\nabla F^1(\mathbf{x}_0)|$, which must of course be $\mathbf{0}$ when evaluated at $\mathbf{x}_1$. Box *et al* choose r_1 as the maximum power of 10 for which $\nabla F^1(\mathbf{x}_0)^T \nabla f(\mathbf{x}_0) \geqslant 0$, so that the gradient directions of f and F^1 at $\mathbf{x}_0$ form an acute angle and $F^1(\mathbf{x}_1) \leqslant f(\mathbf{x}_0)$.

3. Compute the unconstrained minimum of F^k for the current value of r_k using a suitable algorithm for unconstrained minimization.
Second order methods, for example quasi-Newton methods, are preferable, but the precise choice for a given problem will depend on the order of analytic derivatives available. In their Chapter 8, Fiacco and McCormick discuss the relative merits of several methods for the various possibilities, as

well as indicating how special problem structure can be exploited to reduce computational effort. If an algorithm which takes large tentative steps is used, then a procedure for finding a point within violated inequality constraints in the current search direction must be utilized.

4. Terminate the computations if $\mathbf{x}_k$ is acceptable. Since in the limit the penalty terms must vanish, a natural criterion is to compare their sum to zero. Alternatively, the value of the Lagrangian

$$\phi(\mathbf{x}_k; \mathbf{u}_k, \mathbf{v}_k) = f(\mathbf{x}_k) + \Sigma_{i=1}^{l} u_i^k g_i(\mathbf{x}_k) + \Sigma_{j=1}^{m} v_j^k h_j(\mathbf{x}_k) \tag{5.42}$$

and $f(\mathbf{x}_k)$ can be compared. By complementary slackness they must agree at a constrained minimum. When all the problem functions are convex, it can be shown that $\mathbf{x}_k$ and the $\mathbf{u}_k$ and $\mathbf{v}_k$ given by expressions 5.39 and 5.40 constitute a feasible solution to the problem dual to the problem under consideration. The objective function value 5.42 of this dual problem bounds $f(\mathbf{x}_k)$ from below, see Fiacco and McCormick, Chapters 6 and 8 and Dempster (1975), Chapter 6. Box *et al* simply terminate at k for which $r_k = 10^{-4}$.

5. Determine r_{k+1} and return to Step 3. Fiacco and McCormick report that the reduction factor between successive values of r makes little difference to overall computational effort. Indeed, rapid reduction requires fewer unconstrained minimizations, but individual minimizations converge more slowly for smaller values of r. Box *et al* recommend a reduction factor of 10 at each iteration and report that using a factor of 50 sometimes results in a slight loss in accuracy.

Fiacco and McCormick, Chapter 8, give iterative extrapolation formulae based on approximation of $\mathbf{x}_1, \ldots, \mathbf{x}_{k-1}$ by polynomials in $r_1, \ldots, r_{k-1}$ to estimate both $\mathbf{x}_k$ for Step 3 and

$\mathbf{x}_{min}$ for Step 4. These are based on an analysis of the variation of $|\mathbf{x}_k - \mathbf{x}_{min}|$ with r. When Step 3 uses a quadratically convergent (quasi-) Newton method, convergence to an isolated (local) minimum satisfying conditions 5.14 to 5.17 can be guaranteed. The algorithm then becomes a second order constrained technique in which the successive unconstrained minima $\mathbf{x}_k$ converge linearly at a *constant rate* by the fraction α reducing r_k, i.e. the distance from $\mathbf{x}_k$ to the constrained minimum is reduced by α at $\mathbf{x}_{k+1}$.

Other methods

Since it avoids explicit consideration of motion along the constraint, the penalty function approach has the obvious advantage over other effective methods of relative simplicity. On the other hand, as r and the w_i are adjusted to produce convergence of the unconstrained $\mathbf{x}_k$ to the constrained solution, the Hessian matrix of F^k becomes ill-conditioned. Indeed, from 5.30,

$$\nabla^2 F^k(\mathbf{x}) = \nabla^2 f(\mathbf{x}) + \Sigma_{i=1}^{l} w_i G_1 \nabla^2 g_i(\mathbf{x}) + \Sigma_{j=1}^{m} H_1 \nabla^2 h_j(\mathbf{x})$$
$$+ \Sigma_{i=1}^{l} w_i G_{11} \nabla g_i(\mathbf{x})^{\mathrm{T}} \nabla g_i(\mathbf{x}) + \Sigma_{j=1}^{m} H_{11} \nabla h_j(\mathbf{x})^{\mathrm{T}} \nabla h_j(\mathbf{x}),$$

where G_{11} and H_{11} denote the second partial derivatives of G and H with respect to their first arguments, evaluated at $(\mathbf{x}, r_k)$. As r tends to zero the first three terms of expression 5.43 tend to the Hessian of the Lagrangian function ϕ for the problem by virtue of 5.39 to 5.41. However the last two terms tend to infinity due to the contribution of the scalar factors $w_i G_{11}$ and H_{11}. For example, for G and H given by 5.35 and 5.32, these factors become

$$\frac{6w_i r}{g_i^4(\mathbf{x})} = \frac{-2w_i r}{g_i^3(\mathbf{x})} \cdot \frac{-3}{g_i(\mathbf{x})} = u_i \cdot \frac{-3}{g_i(\mathbf{x})} \quad \text{and} \quad \frac{2}{r}$$

In general, it can be shown (see Lootsma, 1972a) that corresponding to each constraint active (as an equation) at the constrained minimum, the Hessian of F^k has an eigenvalue tending to infinity with r^{-1}. The remaining eigenvalues tend to a finite positive limit. The *condition number* of the Hessian, i.e. the ratio of its largest to its smallest eigenvalue, thus tends to infinity with r^{-1}. Since $r_{k+1} < r_k$, and thus for example

$$\frac{6w_i r_{k+1}}{g_i^4(\mathbf{x}_k)} < \frac{6w_i r_k}{g_i^4(\mathbf{x}_k)} \quad \text{and} \quad \frac{2}{r_{k+1}} > \frac{2}{r_k},$$

it is obvious that the contribution from G at the beginning of the $(k+1)$st unconstrained minimization at least initially reduces the eigenvalues of the Hessian of F^{k+1} from those of the Hessian of F^k. The contribution of H does the reverse. For this reason the problem is not so serious with interior point as with exterior point and mixed penalty functions. This is particularly true when the extrapolation formulae mentioned above are used to estimate minima very near constraint boundaries. Since acceleration of convergence of most unconstrained techniques depends on the accuracy of a positive definite quadratic approximation to the function being minimized, we can expect difficulty when the Hessian matrix is ill-conditioned and thus changing rapidly. Further, linear search by quadratic approximation is inappropriate. A second-order method with frequent up-dating of the inverse Hessian will of course ameliorate the difficulties, but usually the price paid is the specification of analytic derivatives.

In order to overcome these difficulties, while retaining linear convergence of successive unconstrained minima at an arbitrary constant rate, Powell (1969) proposed a generalized global penalty function for the equation constrained problem of the form

$$H(\mathbf{x}, \mathbf{r}, \mathbf{s}) = \Sigma_{j=1}^{m} r_j [s_j - h_j(\mathbf{x})]^2, \qquad (5.44)$$

where $\mathbf{r}$ and $\mathbf{s}$ are m vectors of parameters adjusted at each unconstrained minimization. (Note that the terms of 5.44 reduce to the quadratic loss function 5.32 if $r_j = r^{-1}$ and $s_j = 0$ for $j = 1, \ldots, m$.) Define $H^k(\mathbf{x}) = H(\mathbf{x}, \mathbf{r}_k, \mathbf{s}_k)$ and let $\mathbf{x}_k$ be the unconstrained minimum of $F^k = f + H^k$. Then it is clear that $\mathbf{x}_k$ is a solution of the equation constrained problem of minimizing f subject to

$$h_j(\mathbf{x}) = h_j(\mathbf{x}_k) \quad j = 1, \ldots, m,$$

for otherwise $H^k(\mathbf{x}_k)$ could not be minimal. It follows that to solve the original problem one needs an efficient scheme for adjusting the parameter vectors **r** and **s** so as to steer the sequences $h_j(\mathbf{x}_k)$, $j = 1, \ldots, m$, to zero. This can be effected by keeping r constant and changing **s** according to

$$\mathbf{s}_{k+1} = \mathbf{s}_k - \mathbf{h}(\mathbf{x}_k), \tag{5.45}$$

where **h** is the m vector-valued function whose coordinates are the m constant functions h_j. Powell has shown that up to terms of order r_j^{-1}, 5.45 is approximately the Gauss–Newton iteration (see 3.31) for solving by least squares the system of constraint equations as functions of s given by $h_j[x_k(\mathbf{s})]$. As a consequence he proves that for the r_j sufficiently large, $\eta_k = \max_j |h_j(x_k)|$ converges linearly to 0 at as fast a rate as is required. In practice, scaling the functions conformably so that $r_j = 1$, $j = 1, \ldots, m$ is recommended, and the iteration 4.45 is used as long as

$$\eta_{k+1} \leqslant \tfrac{1}{4}\eta_k .$$

When this condition fails, $\mathbf{r}_{k+1}$ is set to $10\mathbf{r}_k$ and $\mathbf{s}_{k+1} = 10^{-1}\mathbf{s}_k$, but this device is seldom required and the unconstrained minimizations usually quickly become trivial.

Since the shift parameters **s** do not materially affect the argument given above for the standard penalty function 5.30, it is easily seen that the dual variables given by

$$v_j^k = 2r_j^k(h_j(\mathbf{x}_k) - s_j^k) \qquad j = 1, \ldots, m$$

converge to the optimal Lagrange multipliers $\bar{v}_j$, providing the Kuhn–Tucker conditions 5.7 hold at the constrained minimum. Otherwise, they diverge together to infinity. (Although Powell has only established the speed of convergence under our regularity condition – when this cannot happen – the result is valid more generally.)

Osborne and Ryan (1972) describe a hybrid algorithm for the general problem (with both inequality and equation constraints) in which they use barrier functions to handle inequality constraints and Powell's penalty function 5.44 to handle equation constraints. Inequality constraints are classified at successive minimizations as active or inactive at the constrained minimum using the current estimates of the dual variables and the values of the constraint functions. The idea is that once all inequality constraints have been classified, the desirable properties and rapid convergence of Powell's method will be enjoyed. The numerical results appear promising.

In general, sequential unconstrained minimization techniques, together with the large step gradient and Lagrangian methods to be described presently, constitute the most effective current methods for constrained optimization.

5.7 Large step gradient methods

Large step gradient methods guide the search for a constrained minimum by moving along the non-linear constraint surface. They utilize a two step procedure to take account of its curvature. First a direction of search from the current point is computed using linearizations, i.e. gradients, of the problem functions and a step is made to a suitable point in what is usually an infeasible direction (particularly if the feasible region is convex). The second step produces a new point as the feasible point nearest, in a suitable sense, to the infeasible point generated at the first step.

Linear equation constraints

The basic strategies and fundamental procedures of large step gradient methods are best introduced with respect to linearly constrained problems. These retain the principal difficulties of the general problem, which occur at step one, while obviating the necessity for step two, which is required by the curvature of the constraint surface. Consider first $s(\leqslant n)$ linearly independent equation constraints of the form

$$\Sigma_{j=1}^{n} k_{rj} x_j = b_r \qquad r = 1, \ldots, s. \tag{5.44}$$

In vector form these constraint equations become

$$\mathbf{K}\mathbf{x} = \mathbf{b}, \tag{5.45}$$

where $\mathbf{K}$ is an $s \times n$ matrix whose rows are n vectors $\mathbf{k}_r^T = (k_{r1}, \ldots, k_{rn})$ and $\mathbf{b}$ is an s vector formed from the right hand sides of 5.44.

In the introduction to this chapter it was pointed out that the equation constraints 5.45 could be used to eliminate s of the n variables using the partition (cf. 5.3)

$$\mathbf{K}_1\mathbf{y} + \mathbf{K}_2\mathbf{z} = \mathbf{b}$$

involving an $s \times s$ non-singular matrix $\mathbf{K}_2$. Since

$$\mathbf{z} = \mathbf{K}_2^{-1}[\mathbf{b} - \mathbf{K}_1\mathbf{y}], \tag{5.46}$$

the objective function f can then be expressed in terms of the $n - s$ independent variables $\mathbf{y}$ as

$$F(\mathbf{y}) = f(\mathbf{y}, \mathbf{K}_2^{-1}[\mathbf{b} - \mathbf{K}_1\mathbf{y}])$$

and, assuming f is twice continuously differentiable, the gradient and Hessian of F in terms of $\mathbf{y}$ are easily evaluated. Indeed, the first order variation of f is given by

$$\nabla_y f(\mathbf{x})\mathbf{y} + \nabla_z f(\mathbf{x})\mathbf{z},$$

where $\nabla_y f(\mathbf{x})$ and $\nabla_z f(\mathbf{x})$ denote the row vectors of first partial derivatives of f with respect to the entries of $\mathbf{y}$ and $\mathbf{z}$ respectively, evaluated at $\mathbf{x}$. The first order variation of f, and hence F, with respect to the independent variables is given by

$$[\nabla_y f(\mathbf{x}) - \nabla_z f(\mathbf{x})\mathbf{K}_2^{-1}\mathbf{K}_1]\mathbf{y} + \nabla_z f(\mathbf{x})\mathbf{K}_2^{-1}\mathbf{b},$$

and thus the gradient of F with respect to $\mathbf{y}$ is

$$\nabla_y f(\mathbf{x}) - \nabla_z f(\mathbf{x})\mathbf{K}_2^{-1}\mathbf{K}_1. \tag{5.47}$$

The Jacobian gradient vector $\mathbf{g}_F$ of F is the transpose of 5.47. Differentiating 5.47 with respect to the elements of $\mathbf{y}$ yields the $(n - s) \times (n - s)$ Hessian matrix of F as

$$\mathbf{H}_{yy} - \mathbf{H}_{yz}\mathbf{K}_2^{-1}\mathbf{K}_1, \tag{5.48}$$

where the Hessian matrix $\mathbf{H}$ of f is partitioned conformably with the partition of $\mathbf{x}$ as

$$\mathbf{H} = \begin{bmatrix} \mathbf{H}_{yy} & \mathbf{H}_{yz} \\ \mathbf{H}_{yz}^{\mathrm{T}} & \mathbf{H}_{zz} \end{bmatrix}. \tag{5.49}$$

In the *elimination of variables* approach to linear equation constrained problems, expressions 5.47 and 5.48 are used directly in a Newton or quasi-Newton algorithm (see for example McCormick, 1970) in $n - s$ variables.

Expression 5.47 was termed the *reduced gradient* of f by Wolfe (1963). (For a more general reduced gradient expression see Fletcher (1972a).) As we have seen, it determines the first order variation of f in the $(n - s)$ dimensional hyperplane determined by the s constraints 5.45 in n dimensional space. Its negative transpose is a natural component of the direction of search in a gradient algorithm; and it was first proposed for use with a steepest descent technique in the *reduced gradient* method of Wolfe. At a constrained minimum $\bar{\mathbf{x}}$ the reduced gradient in the independent variables $\mathbf{y}$ must be zero. It follows that expression 5.47 for the reduced gradient, evaluated at $\bar{\mathbf{x}}$, viz,

$$\nabla_y f(\bar{\mathbf{x}}) - \nabla_z f(\bar{\mathbf{x}})\mathbf{K}_2^{-1}\mathbf{K}_1 = \mathbf{0},$$

gives the components corresponding to the independent variables in the equation constrained version 5.6 of the Kuhn–Tucker conditons 5.7. The analogous expression for the *basic*, or dependent, variables $\mathbf{z}$ is given by

$$\nabla_z f(\bar{\mathbf{x}}) - \nabla_z f(\bar{\mathbf{x}})\mathbf{K}_2^{-1}\mathbf{K}_2 = \mathbf{0}.$$

Hence the optimum values of the multipliers corresponding to the s equation constraints are given by the s vector

$$\bar{\mathbf{w}}^{\mathrm{T}} = -\nabla_z f(\bar{\mathbf{x}})\mathbf{K}_2^{-1}$$

(playing the role of $\bar{\mathbf{v}}$ in 5.6). For future reference note that if a current point $\mathbf{x}_k$ is used to estimate the optimal multipliers $\bar{\mathbf{w}}$ as

$$\mathbf{w}_k^{\mathrm{T}} = -\nabla_z f(\mathbf{x}_k)\mathbf{K}_2^{-1} = \bar{\mathbf{w}}^{\mathrm{T}} - [\nabla_z f(\mathbf{x}_k) - \nabla_z f(\bar{\mathbf{x}})]\mathbf{K}_2^{-1}, \tag{5.50}$$

then, assuming f possesses second derivatives such that the moduli of the columns of the Hessian $\mathbf{H}$ are bounded by K on the level set defined by $\mathbf{x}_0$, the estimate is in error by a term of the order of $|\mathbf{x}_k - \bar{\mathbf{x}}|$. Indeed, using the mean value theorem (see, for example, Apostal (1957), Chapter 6)

$$|[\nabla_z f(\mathbf{x}_k) - \nabla_z f(\bar{\mathbf{x}})]\mathbf{K}_2^{-1}| \leqslant |\nabla f(\mathbf{x}_k) - \nabla f(\bar{\mathbf{x}})|\ |\mathbf{K}_2^{-1}|$$
$$\leqslant |\mathbf{K}_2^{-1}|\ |(\mathbf{x}_k - \bar{\mathbf{x}})^{\mathrm{T}}\tilde{\mathbf{H}}| \leqslant (nK|\mathbf{K}_2^{-1}|)\ |\mathbf{x}_k - \bar{\mathbf{x}}|,$$

where $\tilde{\mathbf{H}}$ denotes evaluation of the columns of the Hessian at points on the line between $\mathbf{x}_k$ and $\bar{\mathbf{x}}$.

Another natural idea, first suggested by Rosen (1960), is to generate a search direction in constrained variables from the orthogonal projection of the negative Jacobian gradient vector $-\mathbf{g}$ onto the $(n - s)$ dimensional hyperplane determined by the constraints. This hyperplane may be expressed as the sum of any vector $\bar{\mathbf{x}}$ satisfying the constraints and vectors in the *tangent space* of the constraints determined by the solutions of the s linear homogeneous equations

$$\mathbf{Kx} = \mathbf{0}. \tag{5.51}$$

For the determination of a search direction it therefore suffices to consider orthogonal projection of $-\mathbf{g}$ onto the tangent space of the constraints. Now it is a result of linear algebra (see, for example, Birkhoff and Maclane, 1953) that n dimensional space may be uniquely decomposed as a vector sum of the solutions of 5.51 and vectors generated by linear combinations of the s transposed rows of the $s \times n$ matrix $\mathbf{K}$. Further, these two sub-spaces are mutually orthogonal. As well as satisfying equation 5.51, the orthogonal projection $\mathbf{u}$ of $-\mathbf{g} = \nabla f(\mathbf{x})^{\mathrm{T}}$ onto the tangent space must therefore also satisfy the equation

$$\mathbf{u} = -\mathbf{g} - \mathbf{K}^{\mathrm{T}}\mathbf{w} \tag{5.52}$$

for some s vector $\mathbf{w}$. Under our assumption that the equations 5.44 are linearly independent, $\mathbf{KK}^{\mathrm{T}}$ is an invertible $s \times s$ matrix. Hence it follows, substituting 5.52 in 5.51, that

$$\mathbf{w} = -(\mathbf{KK}^{\mathrm{T}})^{-1}\mathbf{Kg}, \tag{5.53}$$

and so the negative *projected gradient* is given by

$$\begin{aligned} \mathbf{u} &= [\mathbf{I} - \mathbf{K}^{\mathrm{T}}(\mathbf{K}\mathbf{K}^{\mathrm{T}})^{-1}\mathbf{K}]\,(-\mathbf{g}) \\ &= [\mathbf{I} - \mathbf{P}]\,(-\mathbf{g}), \end{aligned} \tag{5.54}$$

where $\mathbf{I}$ and $\mathbf{P}$ are $n \times n$ matrices, respectively the identity, and an orthogonal projection of rank s. The projection $\mathbf{P}$ may be expressed as $\mathbf{K}^{+}\mathbf{K}$ involving the (*full rank*) *generalized inverse* of the $s \times s$ matrix $\mathbf{K}$ given by the $n \times s$ matrix $\mathbf{K}^{+} = \mathbf{K}^{\mathrm{T}}(\mathbf{K}\mathbf{K}^{\mathrm{T}})^{-1}$. The matrix $\mathbf{P}$ projects vectors onto the space spanned by the transposed rows of $\mathbf{K}$. Since the ith row of $\mathbf{K}$ is the gradient of the ith constraint, these vectors are the normals of the constraint hyperplanes. Hence its complementary projection $\mathbf{I} - \mathbf{P}$ projects vectors so that they are orthogonal to these normals, i.e. so that they lie in the tangent space of the constraints. For a point $\bar{\mathbf{x}}$ to be a constrained minimum, the function f cannot be decreased by search from $\bar{\mathbf{x}}$ in any direction in the tangent space to the constraints. A necessary condition for $\bar{\mathbf{x}}$ to be a constraint minimum is therefore that the projected gradient, i.e. the component $-\mathbf{u}$ of $\mathbf{g}$ in the tangent space, is zero. From 5.52 and 5.53 this may be expressed as

$$\nabla f(\bar{\mathbf{x}}) + \bar{\mathbf{w}}^{\mathrm{T}}\mathbf{K} = \mathbf{0}, \tag{5.55}$$

where

$$\bar{\mathbf{w}}^{\mathrm{T}} = -\bar{\mathbf{g}}^{\mathrm{T}}\mathbf{K}^{+} = -\nabla f(\bar{\mathbf{x}})\mathbf{K}^{+}. \tag{5.56}$$

But expression 5.55 is just the Kuhn–Tucker condition 5.6 for the equation constrained case, with the vector of optimal multipliers $\bar{\mathbf{v}}$ given by $\bar{\mathbf{w}}$. A similar argument to that given above for the estimate 5.50, shows that when the optimal multipliers are estimated at $\mathbf{x}_k$ by

$$\mathbf{w}_k^{\mathrm{T}} = -\nabla f(\mathbf{x}_k)\mathbf{K}^{+} \tag{5.57}$$

the error of the estimate is again of the order of $|\mathbf{x}_k - \bar{\mathbf{x}}|$.

As mentioned above, the original use of both reduced and projected gradients was as search directions in methods of steepest descent. Second order versions of such algorithms can be obtained using a conjugate gradient technique such as that of Fletcher and Reeves described in Section 4.4. It can be shown (see Fletcher, 1972) that the projected gradient technique is also an elimination of variables approach utilizing a general set of s linear equations which are left implicit in the calculation of the projected gradient. As a consequence, when the Fletcher–Reeves algorithm is used with either projected or reduced gradients, quadratic termination in $n - s$ iterations, as opposed to n iterations in the unconstrained case, is guaranteed. This should be taken into account when resetting search directions.

An alternative approach to such use of second order considerations is to use second derivative information, either directly, or approximately through the use of quasi-Newton techniques. We have already seen how this may be done in the elimination of variables approach to the reduced gradient strategy, but it is of interest in the consideration of inequality constraints to obtain the corresponding second order expression for the Lagrange multipliers. To this end consider the constrained minimization of the second order approximation of f about a given point $\bar{\mathbf{x}}$ satisfying the constraints 5.45. In terms of the increment $\mathbf{\Delta x} = \bar{\mathbf{x}} - \mathbf{x}$ to a constrained minimum we wish to minimize

$$f(\bar{\mathbf{x}}) + \nabla f(\bar{\mathbf{x}})\mathbf{\Delta x} + \tfrac{1}{2}\mathbf{\Delta x}^{\mathrm{T}}\mathbf{H}\,\mathbf{\Delta x} \tag{5.58}$$

subject to

$$\mathbf{K}\,\mathbf{\Delta x} = 0. \tag{5.59}$$

Forming the Lagrangian function given by

$$f(\bar{\mathbf{x}}) + \nabla f(\bar{\mathbf{x}})\,\mathbf{\Delta x} + \tfrac{1}{2}\,\mathbf{\Delta x}^{\mathrm{T}}\mathbf{H}\,\mathbf{\Delta x} + \mathbf{w}^{\mathrm{T}}\mathbf{K}\,\mathbf{\Delta x},$$

and writing down the Kuhn–Tucker condition 5.6 for this problem yields the linear equation system

$$\nabla f(\bar{\mathbf{x}}) + \mathbf{\Delta x}^{\mathrm{T}}\mathbf{H} + \bar{\mathbf{w}}^{\mathrm{T}}\mathbf{K} = 0. \tag{5.60}$$

Together with 5.59, the equations 5.60 can be solved for the (second order) optimal $\mathbf{\Delta x}$ and $\bar{\mathbf{w}}$. Partitioning the two sets of equations in terms of $\mathbf{\Delta y}$ and $\mathbf{\Delta z}$ as before gives $\mathbf{\Delta z} = -\mathbf{K}_2^{-1}\mathbf{K}_1\,\mathbf{\Delta y}$ and

$$\nabla_z f(\bar{\mathbf{x}}) + \mathbf{\Delta y}^{\mathrm{T}}\mathbf{H}_{yz} + \mathbf{\Delta z}^{\mathrm{T}}\mathbf{H}_{zz} + \bar{\mathbf{w}}^{\mathrm{T}}\mathbf{K}_2 = 0$$

for the last s equations of 5.60. (Here H is partitioned according to 5.49.) Substituting for $\mathbf{\Delta z}$ and solving for $\bar{\mathbf{w}}^T$ we have an expression for the optimal multipliers

$$\bar{\mathbf{w}}^T = -\nabla_z f(\bar{\mathbf{x}})\mathbf{K}_2^{-1} - \mathbf{\Delta y}^T[\mathbf{H}_{yz} - \mathbf{K}_1^T\mathbf{K}_2^{-1T}\mathbf{H}_{zz}]\mathbf{K}_2^{-1} \quad (5.61)$$

in terms of $\mathbf{\Delta y}$, the optimal increment in the independent variable $\mathbf{y}$. Since $\mathbf{y}$ is unconstrained, $\mathbf{\Delta y}$ is determined by a Newton step, using 5.47 and 5.48, independently of the constraints, and 5.61 gives a second order expression for the optimal multipliers. In practice, when a Newton or quasi-Newton step is used to determine $\mathbf{y}_{k+1}$ – and hence $\mathbf{z}_{k+1}$ – from $\mathbf{y}_k$, the current second order estimate of the optimal multipliers is given from 5.61 as

$$\mathbf{w}_{k+1}^T = -\nabla_z f(\mathbf{x}_{k+1})\mathbf{K}_2^{-1} - \mathbf{\Delta y}_k^T[\mathbf{H}_{yz}^{k+1} - \mathbf{K}_1^T\mathbf{K}_2^{-1}\mathbf{H}_{zz}^{k+1}]\mathbf{K}_2^{-1}. \quad (5.62)$$

A similar calculation to that employed above for first order estimates shows that this estimate is still in error by a term of the order of $|\mathbf{x}_k - \bar{\mathbf{x}}|$.

Equations 5.59 and 5.50 may be rearranged in vector form to give the linear system

$$\begin{bmatrix} \mathbf{H} & \mathbf{K}^T \\ \mathbf{K} & 0 \end{bmatrix} \begin{bmatrix} \mathbf{\Delta x} \\ \bar{\mathbf{w}} \end{bmatrix} = \begin{bmatrix} -\mathbf{g} \\ 0 \end{bmatrix}, \quad (5.63)$$

where $\mathbf{g}$ as usual denotes the Jacobian gradient vector of f evaluated at $\mathbf{x}$. These equations are solved directly for $\mathbf{\Delta x}$ and $\bar{\mathbf{w}}$ in the second order version of the gradient projection approach. Making use of an easily verified formula for the inverse of the partitioned matrix in 5.63, namely,

$$\begin{bmatrix} \mathbf{S} & \mathbf{K}^{*T} \\ \mathbf{K}^* & -(\mathbf{K}\mathbf{H}^{-1}\mathbf{K}^T)^{-1} \end{bmatrix}$$

where

$$\mathbf{K}^* = (\mathbf{K}\mathbf{H}^{-1}\mathbf{K}^T)^{-1}\mathbf{K}\mathbf{H}^{-1} \quad (5.64)$$

and

$$\mathbf{S} = \mathbf{H}^{-1}[\mathbf{I} - \mathbf{K}^T(\mathbf{K}\mathbf{H}^{-1}\mathbf{K}^T)^{-1}\mathbf{K}\mathbf{H}^{-1}] \quad (5.65)$$
$$= \mathbf{H}^{-1}[\mathbf{I} - \mathbf{K}^T\mathbf{K}^*],$$

it follows that

$$\Delta x = -\mathbf{S}\mathbf{g} \tag{5.66}$$

and

$$\overline{\mathbf{w}} = -\mathbf{K}^*\mathbf{g}. \tag{5.67}$$

It is easily verified that $\mathbf{S}^2 = \mathbf{S}$ and $\mathbf{S}^T = \mathbf{S}$ so that $\mathbf{S}$ is an orthogonal projection. Analogous to the 'metric correcting' interpretation of the Newton increment for unconstrained optimization presented in Section 4.4, these formulae can be considered to effect gradient projection in a metric determined by $\mathbf{H}$ rather than the $n \times n$ identity $\mathbf{I}$. (Indeed, if $\mathbf{H}^{-1}$ is replaced by $\mathbf{I}$ in expressions 5.64 through 5.67 then $\mathbf{K}^* = \mathbf{K}^{+T}$, $\mathbf{S} = \mathbf{I} - \mathbf{P}$ and the standard gradient projection formulae, 5.54 and 5.56 respectively, are recaptured.) The matrix $\mathbf{S}$ is the second order curvature correction to make the contours of f spherical in the $(n - s)$ dimensional hyperplane determined by the constraints; it is of rank $n - s$.

The formulae 4.66 and 4.67 have been made the basis of an algorithm for *quadratic programming*, i.e. the minimization of a quadratic function subject to linear constraints, by Fletcher (1971a). More generally they have been used in methods for minimizing a general function subject to linear equality constraints by Goldfarb (1969) and Murtagh and Sargent (1969). Goldfarb's method relies on the fact that when the DFP algorithm of Section 4.4 (with linear search) is applied to a quadratic f beginning with the initial approximation to $\mathbf{S}$ of rank $n - s$ given by $\mathbf{S}_1 = \mathbf{I} - \mathbf{P}$, it will terminate in $n - s$ iterations with $\mathbf{S}_{n-s} = \mathbf{S}$. Murtagh and Sargent's Method 2 also makes use of a variable metric technique (without linear search) based on the rank-one inverse Hessian updating formula 4.22. Rather than iteratively approximating the value of $\mathbf{S}$ at the constrained minimum directly, they build up the current approximation $\mathbf{S}_k$ for the iteration $\Delta\mathbf{x}_k = \lambda_k \mathbf{S}_k \mathbf{g}_k$ by up-dating estimates of the optimal $\mathbf{H}^{-1}$ and $(\mathbf{K}\mathbf{H}^{-1}\mathbf{K}^T)^{-1}$ for use in 5.65. The current approximation to $\mathbf{H}^{-1}$ is up-dated according to 4.22 and a corresponding rank-one formula may be derived for up-dating the approximation to $(\mathbf{K}\mathbf{H}^{-1}\mathbf{K}^T)^{-1}$. This procedure also allows an estimate of the optimal Lagrange multipliers by using the current estimates for the factors in expression 5.61 for $\mathbf{K}^*$. The resulting algorithm can be shown to converge in $n - s + 1$ iterations for quadratic f and a proof of theoretical convergence to a constrained stationary point of a general f with a bounded Hessian has been established. The algorithm, with appropriate modifications, has also been tested on linear and non-linear inequality constraints, to which we turn.

Constraint handling

Let us now consider the equations 5.45 as the subset of a set of l linear inequality constraints which are *active* at the constrained minimum $\bar{\mathbf{x}}$, i.e. which are the binding form of the $s \leqslant l$ inequality constraints

$$\mathbf{Kx} \leqslant \mathbf{b}.$$

(Here the inequality between vectors is used to denote the requirement that the corresponding ordinary inequality holds for each co-ordinate.) Then in addition to the Kuhn–Tucker condition 5.55, which precludes decrease in f by search from $\bar{\mathbf{x}}$ in the tangent space of the constraints, it is necessary that there does not exist a descent direction $\mathbf{u}$ for f such that for some $\lambda > 0$,

$$\mathbf{K}(\bar{\mathbf{x}} + \lambda\mathbf{u}) \leqslant \mathbf{b}. \tag{5.68}$$

Now feasible directions $\mathbf{u}$ for 5.68 satisfy $\mathbf{Ku} \leqslant \mathbf{0}$, and since

$$\mathbf{KK}^+ = \mathbf{KK}^{\mathrm{T}}(\mathbf{KK}^{\mathrm{T}})^{-1} = \mathbf{I},$$

it follows that feasible directions are generated by non-negative combinations of the s columns of $-\mathbf{K}^+$ as $-\mathbf{K}^+\mathbf{v}$, $\mathbf{v} \geqslant \mathbf{0}$. Hence if $-\bar{\mathbf{g}}$ makes an obtuse angle (see 3.4) with each of the s columns $-\mathbf{v}_r$ of $-\mathbf{K}^+$, i.e. $\bar{\mathbf{g}}^{\mathrm{T}}\mathbf{v}_r \leqslant 0$, then there are no descent directions which are feasible, cf. Figure 21. Put in vector form this is just the requirement that

$$-\bar{\mathbf{g}}^{\mathrm{T}}\mathbf{K}^+ = \bar{\mathbf{w}}^{\mathrm{T}} \geqslant \mathbf{0},$$

i.e. that all the optimal multipliers are non-negative. But this is the Kuhn–Tucker condition 5.9 for the multipliers

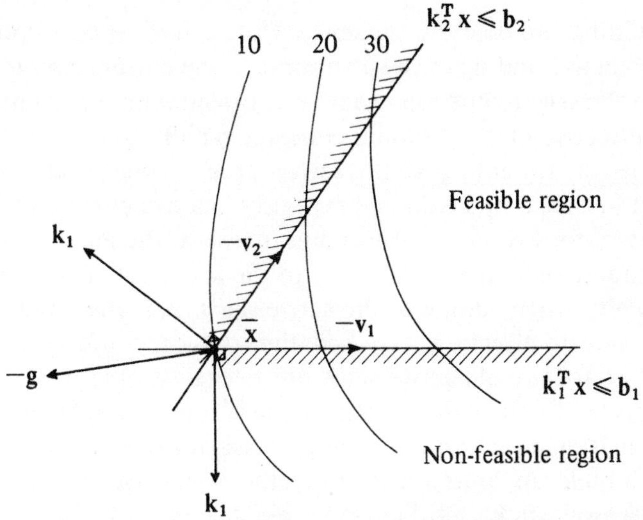

Fig. 21. *The Kuhn–Tucker conditions.*

corresponding to the inequality constraints active at the constrained minimum $\bar{\mathbf{x}}$. Conditions 5.8 and 5.9 are satisfied by setting the remaining multipliers, corresponding to the inequality constraints inactive at $\bar{\mathbf{x}}$, to be zero.

It follows from the above discussion that if

$$w_r = -\mathbf{g}^T \mathbf{v}_r < 0, \tag{5.69}$$

then $\mathbf{v}_r$ is a feasible descent direction and the effect of a movement in that direction is to render the rth constraint inactive, i.e. a strict inequality. This possibility is taken into account in the *active set strategy* for handling inequality constraints. Formally, the *active set* is defined to be the set of indices of the constraints active as equations at the current point. Of course it includes the indices of any equation constraints permanently. With the active set strategy the problem is treated as minimization subject to equation constraints indexed by the current active set. Assuming an initial feasible point, the operations are carried out so that the current point at the end of each iteration remains feasible. Since the initial active set and that corresponding to the constrained minimum will in general differ, rules are required for adding and deleting elements of the active set corresponding to active and inactive constraints at the current point. Once the set of constraints active at the minimum is identified, any effective method should converge rapidly to it.

The rule for adding to the active set is obvious; if at any point in the computation a previously inactive constraint is violated, for example during a linear search, the index of the constraint encountered is added to the active set. The only difficulty when adding to the active set is that when methods of conjugate directions are used, the conjugacy information built up for the old active set is not relevant to the new, and the cycle of conjugate directions must be restarted from the information at the current point. Quasi-Newton methods which build up approximations to the inverse Hessian do not throw away such useful information.

An effective rule for deleting an element from the active set is not so straightforward. A poor choice can have disastrous results for either the total number of function (and derivative) evaluations, or the rate of convergence, or both. As we shall see, the rate of convergence in major iterations can to a certain extent be improved at the expense of more function evaluations at each iteration. The simplest rule is to delete the index r of the rth constraint from the active set if, in light of 5.69, the current (first or second order) estimate of the corresponding Lagrange multiplier w_r is negative. For the purpose of conveniently updating inverse Hessian information, at most a single constraint is often deleted at a single point. Then (in agreement with the usual rules for the simplex method of linear programming) the constraint with the most negative multiplier is rendered inactive. Unfortunately, this simple rule can give rise to a phenomenon known as *zig-zagging*, namely, the oscillation of the active set between two or more sets of integers. (The reader is referred to Zoutendijk (1970) for analytical examples of zig-zagging in which convergence to a point satisfying the Kuhn–Tucker conditions fails, or the rate of convergence of an algorithm with a superlinear rate is rendered linear; see also Luenberger (1972), Chapter 11.) If the constrained minimum for the current active set is found before changes are considered, then a stability check will prevent this set from recurring. Although this is an effective strategy for quadratic programming, it can multiply unnecessary function evaluations for general f, when a necessary change in the active set may be detectable far earlier in the iteration. As a compromise, various *ad hoc* rules have been devised, such as retaining an element of the active set for a number of iterations, fixing an element of the active set upon its re-entry to the set, and only considering deletion of an element from the set if an unconstrained minimum along a line is found in linear search. A more theoretical approach, first proposed for use with the original gradient projection method by Rosen and utilized in the second order gradient projection algorithms of Goldfarb and Murtagh and Sargent, is to compare an estimate of the function decrement achievable with the current active set to that achievable when the index of a suitable constraint is removed (see the original papers or Fletcher (1972) for details). All these rules seem to work well in preventing zig-zagging in practice, but so far few have figured in theoretical proofs that the phenomenon cannot occur. An exception is the multi-iteration retention strategy which is used in a second order algorithm of Ritter (1972) (similar to those of Goldfarb and Murtagh and Sargent) for which superlinear convergence is established. A strategy which is theoretically based, i.e. using it there is an iteration number after which the active set remains constant, has been given by McCormick (1969a, 1970). Constraints are dropped using first order estimates of the current multipliers according to 5.69, but when linear search encounters a constraint it continues in the direction of the projection of the current direction on the constraint until either a minimum along a line is reached, or the active set includes the indices of all constraints. In general however it is important to realize that unless *all* constraints are tested for feasibility at each iteration, even after the constraints active at the constrained minimum have been identified, the minimization process can be led into the infeasible region, see Evans and Gould (1972).

When the active set changes, equation constraints are added

to or deleted from the current problem and the various matrices used by the methods must be up-dated accordingly. The up-dating can generally be effected by a rank one correction to each of the required matrices if at most a single constraint is added or deleted at a time. For example, with the gradient projection method when a constraint $\mathbf{k}_r^{\mathrm{T}}\mathbf{x} = b_r$ is added, and before the next step is calculated, the projection $\mathbf{P}$ is up-dated using the formula

$$\mathbf{P} + (\mathbf{I} - \mathbf{P})\mathbf{k}_r\mathbf{k}_r^{\mathrm{T}}(\mathbf{I} - \mathbf{P})/\mathbf{k}_r^{\mathrm{T}}(\mathbf{I} - \mathbf{P})\mathbf{k}_r$$

for the new projection. In second order versions the projection $\mathbf{S}$ can be up-dated similarly using the formula

$$\mathbf{S} - \mathbf{S}\mathbf{k}_r\mathbf{k}_r^{\mathrm{T}}\mathbf{S}/\mathbf{k}_r^{\mathrm{T}}\mathbf{S}\mathbf{k}_r.$$

Rank one formulae for constraint deletion are somewhat more complicated and the reader is referred to the original papers for details. An advantage of Murtagh and Sargent's method over Goldfarb's, although it is somewhat more expensive in house-keeping operations, is that the former's constraint deletion correction is accurate in terms of second order information, while the latter's only retains enough information to make an approximation in terms of first order information, i.e. appropriate to the case $\mathbf{S} = \mathbf{I} - \mathbf{P}$. With the reduced gradient method it is possible to effect addition or deletion of constraints from the required matrices by the standard pivot step of the simplex method for linear programming. All these rank one recursion formulae effectively reduce the up-dating from order n^3 to n^2 operations, an important saving. They make the possibility of numerical error analysis and effective use of the methods with large sparse constraint structures a reality (cf. Fletcher, 1972).

Another approach to changes in the active set without zigzagging is to determine them automatically using linear or quadratic programming approximations to the original problem. Although a number of such methods have been proposed (and are of considerable use with special assumptions such as convexity, see for example Fletcher, Wolfe (1967) or any of the texts previously referenced), most rely on an *ad hoc* rule to prevent zig-zagging after all. Fletcher (1970b) has, however, given an effective method for minimizing a general function subject to linear inequality constraints. At each iteration it solves the quadratic programme with objective function given by 5.58 and constraints $\mathbf{G}\,\boldsymbol{\Delta}\mathbf{x} \leqslant \mathbf{0}$, where $\mathbf{G}$ is the $l \times n$ constraint matrix of the inequality constraints, subject to a bound on the magnitude of the maximum Δx_i. This bound is adjusted at each iteration so that the resulting feasible hypercube around the current point is the largest possible over which the quadratic approximation to f is accurate to a prescribed degree. Convergence is established both when the actual Hessian $\mathbf{H}$ is used in the approximation to f and when it is estimated by a matrix $\mathbf{T}$ updated according to the DFP formula of Section 4.4 (which corresponds to updating an approximation $\mathbf{S}$ to $\mathbf{H}^{-1}$ with the BFS formula). Obviously the updating of $\mathbf{T}$ depends only on $\boldsymbol{\Delta}\mathbf{x}$ and $\boldsymbol{\Delta}\mathbf{g}$ and not on constraint information, so that no resetting is necessary. On the other hand, the solution of a quadratic programme is of the order of n^3 operations as opposed to the order of n^2 operations per iteration required by quasi-Newton algorithms using the active set strategy. The matrix $\mathbf{T}$ need only be maintained bounded in modulus (as an n^2 vector of its elements) to establish the impossibility of zig-zagging. A similar approach to extending the Levenberg–Marquardt algorithm of Section 4.4 to solve linearly constrained least squares problems has been given by Schrager (1970).

The simplest approach to inequality constraints is simply to convert all inequalities $\mathbf{g}_i^{\mathrm{T}}\mathbf{x} \leqslant 0$ to equation constraints of the form $\mathbf{g}_i^{\mathrm{T}}\mathbf{x} - s_i = 0$ involving non-negative variables s_i, $i = 1, \ldots, l$. Although the number of variables is thus raised by l, the effect of the constraints is to reduce the number of independent variables to n, the number in the unconstrained problem. Moreover, the slack variables are easily evaluated and do not effect the value of f or the g_i; they must be evaluated with the active set strategy anyhow. With slack variables, projection or transformation operators are computed with the $m \times n$ matrix $\mathbf{G}$, rather than the $s \times n$ matrix $\mathbf{K}$ (where s is generally considerably less than n) utilized in the active set strategy. In the linear case however these computations need be made only once rather than iterated at a cost of the order of n^2 operations per iteration. In the non-linearly constrained case, the effects of larger matrices will depend on the method. At least some evidence exists which suggests that the use of slack variables is an effective method for handling inequality constraints, both linear and non-linear.

Gradient projection

The earliest implementation of the two step procedure for handling the curvature of non-linear constraints was given by Rosen (1961) for the gradient projection method. It projects the negative objective function gradient $-\mathbf{g}$ onto the hyperplane tangent to the constraints at the current point to obtain a direction for a linear search to a point minimizing f. It then finds a point on the constraint surface nearest in Euclidean distance to this point at which the value E of f is reduced, see Figure 22. As in the linear case, current practice is to

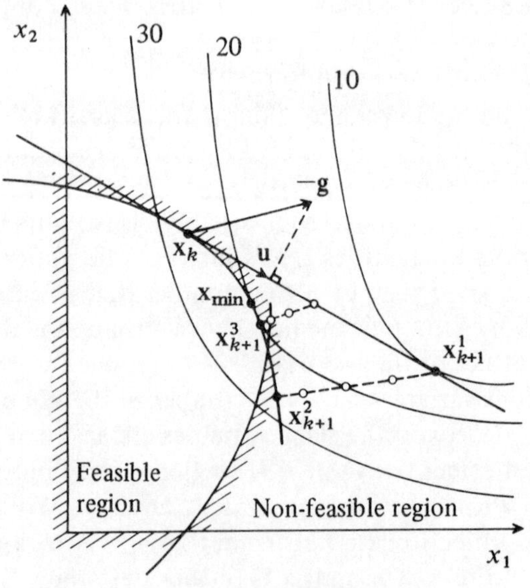

Fig. 22. *A gradient projection iteration.*

employ a second order search direction generated by a conjugate gradient or quasi-Newton algorithm applied to the projected gradient. As mentioned above, the difficulty with the former is that the cycle of conjugate directions must be reset each time the active set changes.

Consider the equation constraints

$$k_r(\mathbf{x}_k) = 0 \qquad r = 1, \ldots, s \leqslant n. \tag{5.70}$$

Their *tangent space* at a point $\mathbf{x}_k$ is given by the solutions $\mathbf{x}$ of the linear homogenous equations

$$\nabla k_r(\mathbf{x}_k)\mathbf{x} = 0 \qquad r = 1, \ldots, s.$$

In vector form these equations become

$$\mathbf{Kx} = \mathbf{0}, \tag{5.71}$$

where $\mathbf{K}$ is the $s \times n$ matrix of partial derivatives of the s functions k_r evaluated at $\mathbf{x}_k$.

At each iteration the linearization 5.71 of the non-linear constraints is used to compute the negative projected gradient from 5.54 as

$$\mathbf{u} = [\mathbf{I} - \mathbf{K}^T(\mathbf{KK}^T)^{-1}\mathbf{K}](-\mathbf{g}),$$

In outline step by step form, the gradient projection method for minimizing f subject to inequality and equality constraints, 5.1 and 5.2, proceeds as follows.

Step: 1. Find an initial point $\mathbf{x}_0$ in the feasible region, i.e. satisfying all constraints $g_i(\mathbf{x}_0) \leqslant 0, i = 1, \ldots, l$ and $h_j(\mathbf{x}_0) = 0, j = 1, \ldots, m$, using any suitable method.

2. Form the current set of equation constraints 5.70 from the equation constraints 5.2 and the currently binding inequality constraints 5.1, i.e. those for which $g_i(\mathbf{x}_k) \geqslant 0$, and compute $\mathbf{K}$. Note that under our regularity assumption the number of elements in the current active set, which indexes the current set of quasi-Newton constraints, does not exceed n.

3. Compute the current search direction $\mathbf{u}$ from the negative projected gradient using a suitable method.

4. Perform a linear search to minimize $f(\mathbf{x}_k + \lambda\mathbf{u})$ at $\mathbf{x}_{k+1}^1 = \mathbf{x}_k + \lambda_{k+1}\mathbf{u}$. Generally $\mathbf{x}_{k+1}^1$ is infeasible. If $\mathbf{x}_{k+1}^1$ *is* feasible set $\mathbf{x}_{k+1}^2 = \mathbf{x}_{k+1}^1$ and go to Step 6.
5. Apply a suitable version of Newton's method to the non-linear equations 5.44 to attempt to find a point $\mathbf{x}_{k+1}^2$ on the boundary of the feasible region satisfying the current set of equation constraints.
6. Compare $f(\mathbf{x}_k)$ and $f(\mathbf{x}_{k+1}^2)$. If $f(\mathbf{x}_{k+1}^2) < f(\mathbf{x}_k)$, set $\mathbf{x}_{k+1} = \mathbf{x}_{k+1}^2$ and return to Step 2. Otherwise repeat Steps 5 and 6, successively halving λ_{k+1} to generate a sequence of points $\mathbf{x}_{k+1}^t = \mathbf{x}_k + 2^{-t+2}\lambda_{k+1}\mathbf{u}$, until for some $t > 2$, $\mathbf{x}_{k+1} = \mathbf{x}_{k+1}^t$ is found.
7. Iterate Steps 2 to 6 until the constrained minimum $\bar{\mathbf{x}}$ is located. The optimal multipliers, i.e. the $\bar{u}_i$ corresponding to binding inequality constraints and $\bar{\mathbf{v}}$, can then be computed from 5.53 as $\bar{\mathbf{w}} = -(\bar{\mathbf{K}}\bar{\mathbf{K}}^{\mathrm{T}})^{-1}\bar{\mathbf{K}}\mathbf{g}$.

The non-linearly constrained versions of Goldfarb's and Murtagh and Sargent's methods also make use of linearization of the constraints about the current point but do not take account of their curvature with the strategy embodied in Steps 4 and 5. Instead, as a non-linear constraint is encountered or relaxed its linearization is added to or deleted from the active set. At each point the required matrices must be updated for the current linearizations. The rank one recursion formulae developed for linear constraints are used. In practice, gradient projection methods have not proved as effective as the reduced gradient method to which we now turn.

The reduced gradient method

Consider the current set of s equality constraints 5.70 and suppose that the vector $\mathbf{x}_k$ is divided into two components as

$$\mathbf{x}_k^{\mathrm{T}} = (\mathbf{y}_k \, \mathbf{z}_k), \tag{5.72}$$

where $\mathbf{y}_k$ is an $n - s$ vector of *independent* variables and $\mathbf{z}_k$ is an s vector of *basic* variables. Then if the constraints are continuously differentiable, the implicit function theorem guarantees the existence of a function **G** such that for all **y** in a neighbourhood of $\mathbf{y}_k$,

$$\mathbf{z} = \mathbf{G}(\mathbf{y})$$

$$k_r(\mathbf{y}, \mathbf{G}(\mathbf{y})) = 0 \qquad r = 1, \ldots, s,$$

and **G** is continuously differentiable, see Apostal (1957), Chapter 7. Partitioning the equations 5.71 defining the tangent space of the constraints at $\mathbf{x}_k$ correspondingly yields

$$\nabla_y k_r(\mathbf{x}_k)\mathbf{y} + \nabla_z k_r(\mathbf{x}_k)\mathbf{z} = 0 \qquad r = 1, \ldots, s,$$

where $\nabla_y k_r(\mathbf{x}_k)$ and $\nabla_z k_r(\mathbf{x}_k)$ denote the row vectors of first partial derivatives of k_r with respect to the entries of **y** and **z** respectively, evaluated at $\mathbf{x}_k$. In vector form these constraint linearizations become

$$\mathbf{K}_1\mathbf{y} + \mathbf{K}_2\mathbf{z} = \mathbf{0}.$$

The steps of the *reduced gradient* method, due to Wolfe (1963, 1967) and Abadie and Carpentier (1965), are the same

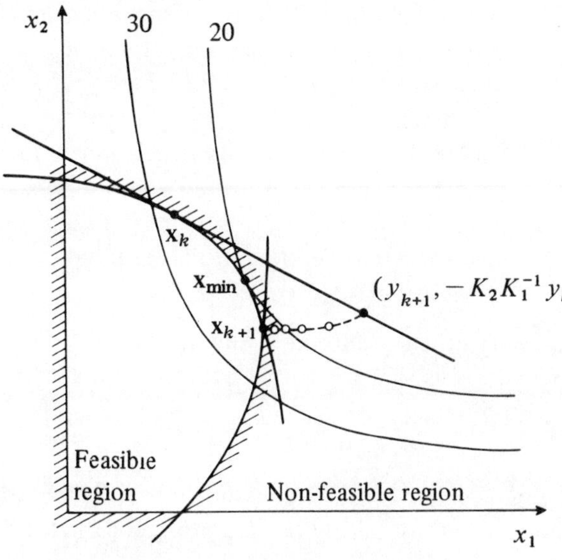

Fig. 23. *A reduced gradient iteration.*

as those of the projected gradient method except for the basic Steps 3 to 5 which become the following, see Figure 23.

Step: 3. Compute $\nabla_z f(\mathbf{x}_k)$ and the current search direction $\mathbf{u}$ for the independent variables $\mathbf{y}$ from the negative reduced gradient using a suitable method.

4. Perform a linear search to minimize $f[\mathbf{y}_k + \lambda\mathbf{u}, -\mathbf{K}_2^{-1}\mathbf{K}_1(\mathbf{y}_k + \lambda\mathbf{u})]$ at a point $\mathbf{y}_{k+1}^1 = \mathbf{y}_k + \lambda_{k+1}\mathbf{u}$.

5. Compute a feasible solution $\mathbf{x}$ to the current set of equation constraints by attempting to solve the system

$$g_r(\mathbf{y}_{k+1}^1, \mathbf{z}) = 0 \qquad r = 1, \ldots, s \tag{5.73}$$

in the remaining s variables $\mathbf{z}_{k+1}^1$ of $\mathbf{x}_{k+1}$ using a suitable version of Newton's method. Abadie and Guignou (1970) recommend the pseudo-Newton iteration

$$\mathbf{z}_{t+1} = \mathbf{z}_t - \mathbf{K}_2^{-1}\boldsymbol{\gamma}(\mathbf{y}_{k+1}^1, \mathbf{z}_t),$$

where the entries of the s vector $\boldsymbol{\gamma}$ are given by the expressions in 5.73. For sufficiently small λ_{k+1} convergence to $\mathbf{z}_{k+1}^1$ is rapid. If convergence is too slow halve λ_{k+1} to generate a new $\mathbf{y}_{k+1}^1 = \mathbf{y}_k + \frac{1}{2}\lambda_{k+1}\mathbf{u}$ and repeat the attempt. The current set of equation constraints is suitably modified as new inequality constraints are encountered and old ones cease to bind. Accordingly, the current partition of $\mathbf{x}$ into basic and independent variables must be modified as s is augmented or reduced.

At a constrained minimum $\bar{\mathbf{x}}$ the optimum values of the multipliers corresponding to the equation constraints are given from 5.50 by

$$\bar{\mathbf{w}} = -\nabla_z f(\bar{\mathbf{x}})\mathbf{K}_2^{-1}(\bar{\mathbf{x}})$$

The Kuhn–Tucker optimality conditions 5.7, 5.8 and 5.9 are satisfied as usual by taking the remaining optimal multipliers to be zero.

Abadie and his associates have coded a version of the reduced gradient method which is probably the most extensively tested and efficient constrained optimization routine at the time of writing. As well as proving the most efficient algorithm applied to an international set of test problems, it has been effective in solving large optimal control problems in discrete time, see Abadie (1969). Abadie's Generalised Reduced Gradient (GRG) 69 code uses slack variables to convert to equations all inequality constraints except a complete set of variable bounds. (The latter can always be assumed present in practice.) Basic and independent variable partition is implemented with respect to these constraints and the linear search of Step 4 is effected by quadratic approximation. The principal advantage of reduced gradient over projected gradient methods probably stems from the fact that at Step 5 the former must solve a set of non-linear equations in s unknowns, while the latter must solve the same set for all n variables of $\mathbf{x}$. When the constraint functions g_i, h_j are linear, one would expect the relative efficiencies of general codes for the two algorithms to be reversed, since the projected gradient method then requires essentially no Step 5, while the reduced gradient method must still solve a set of linear equations in s basic variables. Limited empirical results tend to confirm this expectation (see Abadie (1969), Table 2), but as we have seen, the use of rank one updating formulae makes the two efforts comparable. In terms of major iterations, Luenberger (1972) gives a careful analysis of the rates of convergence of the original (i.e. steepest descent) versions of the gradient projection and reduced gradient algorithms. These are found to converge linearly at the rate of steepest descent (see Section 4.2), but with the condition number of the Hessian $\mathbf{H}$ replaced by the condition numbers of the Hessian of the Lagrangian of the problem evaluated at the constrained minimum in terms, respectively, of the original and the independent variables. Although of the same order, the relation of the speeds of convergence in major iterations of the two algorithms therefore depends on the final choice of independent variables in the reduced gradient method. The reduced gradient convergence rate will be considerably worsened by ill-conditioning of the final basis matrix $\mathbf{K}_2(\bar{\mathbf{x}})$ through a poor choice of independent variables. Luenberger's arguments can be extended to give the theoretical rates of convergence in major iterations of the two methods as quadratic or super-linear when used with appropriate second order (i.e. Newton or quasi-Newton) algorithms.

5.8 Lagrangian methods

Lagrangian methods combine theoretical considerations with features of both large step gradient methods and the global penalty function approach. Current approaches may be divided into three categories: sequential unconstrained techniques, methods which solve a sequence of linearly unconstrained problems, and methods requiring only a single unconstrained minimization of the Lagrangian function through treating the estimates of the multipliers found in the previous section as functions of **x**. To understand the relations between these methods a certain amount of modern optimization *theory* is required. This will be summarized briefly before the methods are outlined.

Local and global duality theory

The theoretical analysis of the general problem of minimizing f subject to non-linear constraints, given in vector form as

$$\mathbf{g}(\mathbf{x}) \leqslant \mathbf{0} \qquad \text{and} \qquad \mathbf{h}(\mathbf{x}) = \mathbf{0},$$

makes use of auxiliary theoretical minimizations. Much of the detail of the theory concerns conditions under which these minima exist and possess certain properties as functions of the parameters involved. However, for our purposes it will be sufficient to assume most of the difficulty away. (The interested reader is referred to Luenberger (1969, 1972) or Dempster (1975) for details.)

For simplicity consider first the problem constrained by m equations under the assumption that the problem functions are twice continuously differentiable. In order to study the existence of Lagrange multipliers one introduces a new function, termed the *primal function* p of the problem, whose argument is an m vector of slack variables s. This is effected by means of auxiliary theoretical minimizations in the definition of p as

$$p(\mathbf{s}) = \text{minimum } \{f(\mathbf{x}) \text{ subject to } h(\mathbf{x}) = \mathbf{s}\}. \quad (5.74)$$

Notice that the definition of the primal function can be considered to embed the original problem in a parametric family of similar problems indexed by the vectors s. At $\mathbf{s} = \mathbf{0}$ the original problem is recaptured and $p(\mathbf{0}) = f_{\min}$. (It is of course possible that $p(\mathbf{s})$ is infinite for some s, but assuming the constrained minimum of the original problem exists, it is finite in a neighbourhood of $\mathbf{0}$ by the continuity of the problem functions.) The behaviour of the *fundamental surface* of points $(\mathbf{s}, p(\mathbf{s}))$ in $(m + 1)$ dimensional space, defined in a neighbourhood of the point $(\mathbf{0}, f_{\min})$ by p, is basic to the existence of Lagrange multipliers and a duality theory for the original problem. Since the problem functions are differentiable, the gradient ∇p of p can be defined in a neighbourhood of $\mathbf{s} = \mathbf{0}$, but may be infinite in all co-ordinates at $\mathbf{0}$. It was the purpose of our regularity condition in Section 5.1 to preclude this possibility, and for a locally regular problem it can be shown that ∇p exists in a neighbourhood of $\mathbf{0}$. The equation of the tangent hyperplane to the fundamental surface at the point $(\mathbf{0}, p(\mathbf{0}))$ is given (see Figure 24) by

$$-\nabla p(\mathbf{0})\mathbf{s} + 1t = p(\mathbf{0}) = f_{\min}, \quad (5.75)$$

Expressing the point $(\mathbf{0}, f_{\min})$ in $(m + 1)$ dimensional space in terms of the constrained minimum $\bar{\mathbf{x}}$ as $(h(\bar{\mathbf{x}}), f(\bar{\mathbf{x}}))$, we therefore have, in particular,

$$f(\bar{\mathbf{x}}) - \nabla p(\mathbf{0})\mathbf{h}(\bar{\mathbf{x}}) = f(\bar{\mathbf{x}}) . \quad (5.76)$$

Denote by $\mathbf{x}(\mathbf{s})$ the minimizing point for the problem 5.74 for a slack vector s in the neighbourhood of $\mathbf{0}$. Then points on the fundamental surface near $(\mathbf{0}, p(\mathbf{0}) = (\mathbf{h}(\bar{\mathbf{x}}), f(\bar{\mathbf{x}}))$ can be parametrized by such points $\mathbf{x}(\mathbf{s})$ as $(\mathbf{h}(\mathbf{x}(\mathbf{s})), f(\mathbf{x}(\mathbf{s})))$. Since the hyperplane with equation 5.75 is tangent to the fundamental surface at $(\mathbf{h}(\bar{\mathbf{x}}), f(\bar{\mathbf{x}}))$, it follows (see Figure 24)

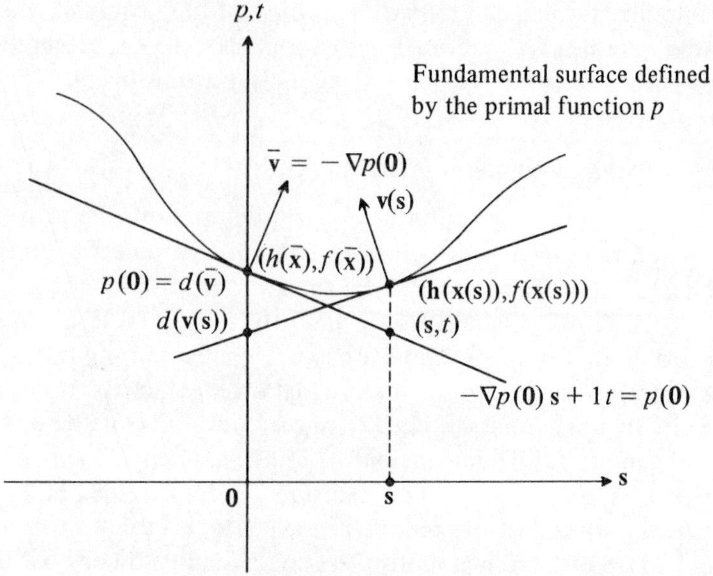

Fig. 24. *Fundamental surface of a locally convex problem.*

that the expression

$$f(\mathbf{x}) - \nabla p(\mathbf{0})\mathbf{h}(\mathbf{x}) - f(\bar{\mathbf{x}}) + \nabla p(\mathbf{0})\mathbf{h}(\bar{\mathbf{x}})$$

must have a stationary point at $\bar{\mathbf{x}}$, i.e.

$$\nabla f(\bar{\mathbf{x}}) - \nabla p(\mathbf{0})\nabla \mathbf{h}(\mathbf{x}) \; = \; \mathbf{0}. \tag{5.77}$$

But 5.77 is again the familiar Kuhn–Tucker condition 5.6, and we have found another expression for the optimal Lagrange multiplier vector $\bar{\mathbf{v}}$ as $-\nabla p(\mathbf{0})^{\mathrm{T}}$. The same argument applies to points on the fundamental surface generated by slack vectors s in a neighbourhood of $\mathbf{0}$ where $p(\mathbf{s})$ exists. This gives an expression for the optimal Lagrange multipliers for the problem 5.74 as $\mathbf{v}(\mathbf{s}) = -\nabla p(\mathbf{s})^{\mathrm{T}}$, the non-trivial co-ordinates of the normal vector to the non-vertical tangent hyperplane to the fundamental surface at the point $(\mathbf{s}, p(\mathbf{s}))$. Hence, $\mathbf{x}(\mathbf{0}) = \bar{\mathbf{x}}$, the constrained minimum, and $\mathbf{v}(\mathbf{0}) = \bar{\mathbf{v}}$, the vector of optimal multipliers.

To define a local dual problem for the original problem, we

fix the multipliers at their optimal value $\bar{\mathbf{v}}$ and consider as a function of $\mathbf{x}$ the Lagrangian of the original problem given by the appropriate version of 5.10 as

$$\phi(\mathbf{x};\bar{\mathbf{v}}) = f(\mathbf{x}) + \bar{\mathbf{v}}^{\mathrm{T}}\mathbf{h}(\mathbf{x}).$$

The first order Kuhn–Tucker conditions 5.77 state that this function has a stationary point at the constrained minimum $\bar{\mathbf{x}}$, with value $f(\bar{\mathbf{x}})$ from 5.76, while the appropriate second order Kuhn–Tucker conditions 5.11 and 5.13 state that at $\bar{\mathbf{x}}$ its Hessian $\mathbf{F}$, given by

$$\mathbf{F}(\mathbf{x}) = \nabla^2 f(\mathbf{x}) + \Sigma_{j=1}^{m} \bar{v}_j \nabla^2 h_j(\mathbf{x}) \tag{5.78}$$

is non-negative definite for vectors in the tangent space to the constraints at $\bar{\mathbf{x}}$. If the $n \times n$ matrix $\mathbf{F}(\bar{\mathbf{x}})$ is positive definite for such vectors, $\bar{\mathbf{x}}$ is an isolated local minimum. We shall henceforth assume that $\mathbf{F}(\bar{\mathbf{x}})$ is positive definite everywhere on n dimensional space. Then the Lagrangian is locally convex in $\mathbf{x}$ near $\bar{\mathbf{x}}$ (see Section 1.4) and has an isolated local minimum at the point $\bar{\mathbf{x}}$, since it satisfies the first and second order conditions for an extremum there. In particular, points $(\mathbf{h}(\mathbf{x}(\mathbf{s})), f(\mathbf{x}(\mathbf{s})))$ on the fundamental surface in a neighbourhood of $(\mathbf{h}(\bar{\mathbf{x}}), f(\bar{\mathbf{x}}))$ must lie above the tangent hyperplane to the surface at this point (see Figure 24). Geometrically, the fundamental surface is convex in a neighbourhood of $(\mathbf{h}(\bar{\mathbf{x}}), f(\bar{\mathbf{x}}))$. (In general the tangent hyperplane at $(\mathbf{h}(\bar{\mathbf{x}}), f(\bar{\mathbf{x}}))$ may cut the fundamental surface in some direction at the point of tangency.)

Now, since by continuity $\mathbf{F}(\mathbf{x}) = \nabla_x^2 \phi(\mathbf{x};\bar{\mathbf{v}})$ is positive definite and hence non-singular near $\bar{\mathbf{x}}$, it follows by the implicit function theorem that the equation

$$\nabla f(\mathbf{x}) + \mathbf{v}^{\mathrm{T}}\nabla\mathbf{h}(\mathbf{x}) = \mathbf{0}$$

has a unique solution $\mathbf{x}$ in a neighbourhood of $\bar{\mathbf{x}}$ when $\mathbf{v}$ is in a neighbourhood of $\bar{\mathbf{v}}$. Hence for fixed $\mathbf{v}$ sufficiently close to $\bar{\mathbf{v}}$, $\phi(\mathbf{x};\mathbf{v}) = f(\mathbf{x}) + \mathbf{v}^{\mathrm{T}}\mathbf{h}(\mathbf{x})$ will have an isolated local minimum in $\mathbf{x}$. But we have already seen that for $\mathbf{s}$ sufficiently close to $\mathbf{0}$, $\mathbf{v}(\mathbf{s}) = -\nabla p(\mathbf{s})^{\mathrm{T}}$ and this minimum is given by the minimum $\mathbf{x}(\mathbf{s})$ of the problem 5.75. Thus in a neighbourhood of $\bar{\mathbf{x}}$ there is a unique continuously differentiable correspondence between $\mathbf{v}$ and $\mathbf{x}$ through the solution of the unconstrained minimization of $\mathrm{f} + \mathbf{v}^{\mathrm{T}}\mathrm{h}$ with respect to $\mathbf{x}$.

This correspondence allows us to dualize the primal parametrization of the fundamental surface of the problem in terms of s, through the dual parametrization of its linearization, $\phi(\mathrm{x}; \mathrm{v}) = f(\mathrm{x}) + \mathbf{v}^{\mathrm{T}}\mathrm{h}(\mathrm{x})$, in terms of $\mathbf{v} = -\nabla p(\mathrm{s})^{\mathrm{T}}$. Near $\mathbf{v}$ one may define the *dual function d* of the problem as

$$d(\mathrm{v}) = \text{minimum } \phi(\mathrm{x}; \mathrm{v}),$$

where it is understood that the minimum is to be taken in a suitable neighbourhood of $\mathrm{x(s)}$ near $x(\mathbf{0}) = \bar{\mathrm{x}}$ and it is achieved at $\mathrm{x(v)} = \mathrm{x(s)}$. Since by definition

$$d(\mathrm{v}) = \phi(\mathrm{x(v)}; \mathrm{v}),$$

we have

$$\begin{aligned} \nabla d(\mathrm{v}) &= [\nabla f(\mathrm{x(v)}) + \mathrm{v}^{\mathrm{T}}\nabla \mathrm{h}(\mathrm{x(v)})]\nabla \mathrm{x(v)} + \mathrm{h}(\mathrm{x(v)})^{\mathrm{T}} \\ &= \mathrm{h}(\mathrm{x(v)})^{\mathrm{T}}. \end{aligned} \tag{5.79}$$

Moreover, differentiating 5.79 gives

$$\nabla^2 d(\mathrm{v}) = \nabla \mathrm{h}(\mathrm{x(v)})\nabla \mathrm{x(v)}, \tag{5.80}$$

and differentiating $\nabla f(\mathrm{x(v)}) + \mathrm{v}^{\mathrm{T}}\nabla \mathrm{h}(\mathrm{x(v)}) = \mathbf{0}$ with respect to $\mathbf{v}$ gives

$$\nabla_x^2 \phi(\mathrm{x(v)}, \mathrm{v})\nabla \mathrm{x(v)} + \nabla \mathrm{h}(\mathrm{x(v)})^{\mathrm{T}} = \mathbf{0}. \tag{5.81}$$

Since $\nabla_x^2 \phi$ is nonsingular, solving 5.81 for the $n \times m$ matrix $\nabla \mathrm{x(v)}$ and substituting in 5.80 yields the Hessian of d as

$$\nabla^2 d(\mathrm{v}) = -\nabla \mathrm{h}(\mathrm{x(v)})[\nabla_x^2 \phi(\mathrm{x(v)}, \mathrm{v})]^{-1}\nabla \mathrm{h}(\mathrm{x(v)})^{\mathrm{T}}, \tag{5.82}$$

which is negative definite near $\bar{\mathrm{v}}$. It follows that the point $\bar{\mathrm{v}}$ corresponding to $\mathrm{x}(\bar{\mathrm{v}}) = \mathrm{x}(\mathbf{0}) = \bar{\mathrm{x}}$ is a local unconstrained maximum of d, since from 5.79,

$$d(\bar{\mathrm{v}}) = \mathrm{h}(\mathrm{x}(\bar{\mathrm{v}}))^{\mathrm{T}} = \mathbf{0}.$$

The unconstrained maximization problem

$$\text{maximize } d(\mathrm{v}) \tag{5.83}$$

is the local *dual problem* for the original problem.

From 5.75 and 5.76,

$$d(\bar{\mathbf{v}}) = \phi(\bar{\mathbf{x}}; \bar{\mathbf{v}}) = f(\bar{\mathbf{x}}) = p(\mathbf{0}).$$

Recall that we have seen that $\mathbf{v} = -\nabla p(\mathbf{s})^{\mathrm{T}}$ in a neighbourhood of $\mathbf{0}$. Dually, it follows from 5.79 that

$$\mathbf{s} = \mathbf{h}(\mathbf{x}(\mathbf{v})) = \nabla d(\mathbf{v})^{\mathrm{T}} \qquad (5.84)$$

in a neighbourhood of $\bar{\mathbf{v}}$.

When the problem functions are convex, the Lagrangian function is a convex function of $\mathbf{x}$, $\bar{\mathbf{x}}$ is a global minimum for the original problem and the fundamental surface lies entirely above the point $(\mathbf{h}(\bar{\mathbf{x}}), f(\bar{\mathbf{x}}))$. It can then be shown that the dual function is concave, so that $\bar{\mathbf{v}}$ is a global maximum of d and $(\bar{\mathbf{x}}, \bar{\mathbf{v}})$ is a saddle-point of the Lagrangian function ϕ. For the general problem, with inequality as well as equation constraints, the entire argument can be extended. In this situation, the primal function is defined by

$$p(\mathbf{s}_2, \mathbf{s}_2) = \text{minimum } \{f(\mathbf{x}) \text{ subject to } \mathbf{g}(\mathbf{x}) \leqslant -\mathbf{s}_1, \mathbf{h}(\mathbf{x}) = \mathbf{s}_2\},$$

the Lagrangian function ϕ is given by 5.10, and the dual function, defined for $\mathbf{u} \geqslant \mathbf{0}$ and $\mathbf{v}$ as

$$d(\mathbf{u}, \mathbf{v}) = \text{minimum } [f(\mathbf{x}) + \mathbf{u}^{\mathrm{T}}\mathbf{g}(\mathbf{x}) + \mathbf{v}^{\mathrm{T}}\mathbf{h}(\mathbf{x})],$$

achieves a local maximum for $\mathbf{u} \geqslant \mathbf{0}$, $\mathbf{v}$ at the optimal multipliers $\bar{\mathbf{u}} \geqslant \mathbf{0}$, $\bar{\mathbf{v}}$ of the Kuhn–Tucker conditions 5.7 to 5.9.

Sequential unconstrained techniques

We are now in a position to amplify the brief discussion of the strategy underlying sequential unconstrained minimization techniques in terms of the duality theory given above. Again for simplicity we restrict initial attention to the equation constrained case. First notice that when the original problem is locally convex (in the sense that the Hessian of the Lagrangian evaluated at the constrained minimum is positive definite) the local dual problem 5.83 can be expressed in terms of *both* $\mathbf{x}$ and $\mathbf{v}$ in the form

$$\text{maximize } \phi(\mathbf{x}, \mathbf{v}) = f(\mathbf{x}) + \mathbf{v}^{\mathrm{T}}\mathbf{h}(\mathbf{x}) \qquad (5.85\text{a})$$

subject to

$$\nabla f(\mathbf{x}) + \mathbf{v}^{\mathrm{T}}\nabla \mathbf{h}(\mathbf{x}) = 0. \qquad (5.85b)$$

Indeed, as we saw above, for $\mathbf{x}$ near $\bar{\mathbf{x}}$ the constraint 5.85b insures that for any $\mathbf{v}$ near $\bar{\mathbf{v}}$ the corresponding $\mathbf{x}(\mathbf{v})$ minimizes the Lagrangian in $\mathbf{x}$; hence $\phi(\mathbf{x}(\mathbf{v}); \mathbf{v}) = d(\mathbf{v})$.

In Section 5.6 we saw that at an unconstrained minimum $\mathbf{x}_k$ of the function F^k, given by

$$F^k(\mathbf{x}) = f(\mathbf{x}) + \Sigma_{j=1}^m H(h_j(\mathbf{x}), r_k),$$

we have the first order condition

$$\nabla F^k(\mathbf{x}_k) = \nabla f(\mathbf{x}_k) + \mathbf{v}_k^{\mathrm{T}}\nabla \mathbf{h}(\mathbf{x}_k) = 0,$$

where the co-ordinates of the multiplier vector $\mathbf{v}_k$ are given by the partial derivatives of the penalty function H as

$$v_j^k = H_1(h_j(\mathbf{x}_k), r_k) \qquad j = 1, \ldots, m.$$

It follows that the sequence of points $\mathbf{x}_k$, $\mathbf{v}_k$ is feasible for the local dual problem 5.84 and under our regularity assumption it converges to $\bar{\mathbf{x}}$, $\bar{\mathbf{v}}$, the constrained minimum and vector of optimum multipliers, respectively. (Note that we can now interpret the simultaneous divergence of the $\mathbf{v}_k$ sequences to $\pm\infty$ as an indication that the hyperplane tangent to the fundamental surface of the problem is vertical at $\mathbf{0}$, so that the Kuhn–Tucker conditions do not hold at $\bar{\mathbf{x}}$.) The sequence $\mathbf{v}_k$ can be thought of as a sequence of evaluations of the dual function of the problem converging to its unconstrained maximum $\bar{\mathbf{v}}$. However the trajectory of these evaluations is determined more or less arbitrarily by the choice of the form of the function F^k (really the penalty function) and the parameter sequence r_k; in particular, no use of derivative information for the dual function is made to guide the trajectory. (The use of extrapolation techniques takes account of this information for F^k, but this is in general the wrong objective function.) Furthermore, we saw that the Hessian of F^k becomes ill-conditioned as $\mathbf{v}_k$ approaches $\bar{\mathbf{v}}$. On the other hand, we have also seen that if $F^k(\mathbf{x})$ is taken to be the Lagrangian $\phi(\mathbf{x}; \mathbf{v}_k)$ itself (as required by 5.85) and an attempt is made to use dual function derivative information to guide the trajectory of multipliers $\mathbf{v}_k$, the Hessian $\mathbf{F}(\mathbf{x})$ of ϕ at $\bar{\mathbf{v}}$ is only necessarily positive definite over the tangent space at $\bar{\mathbf{x}}$. Hence the local duality theory does not necessarily apply. (Fletcher (1970a) gives analytical examples of this phenomenon.) In practice this situation can be awkward, for near to $\bar{\mathbf{x}}$ the minimization procedure may converge very slowly to this stationary point of the Lagrangian or even pass through it.

These difficulties can be overcome by the simple device of replacing the original problem by the minimization of an augmented function of the form

$$f(\mathbf{x}) + \frac{r}{2}\mathbf{h}(\mathbf{x})^{\mathrm{T}}\mathbf{h}(\mathbf{x})$$

subject to $\mathbf{h}(\mathbf{x}) = 0$ for some $r > 0$. In terms of the notation for the Lagrangian of the original problem, the Lagrangian of this problem is given by

$$\phi(\mathbf{x}; \mathbf{v}) + \frac{r}{2}\mathbf{h}(\mathbf{x})^{\mathrm{T}}\mathbf{h}(\mathbf{x}), \tag{5.86}$$

so that its gradient is

$$\nabla_x \phi(\mathbf{x}; \mathbf{v}) + r\mathbf{h}(\mathbf{x})^{\mathrm{T}}\nabla\mathbf{h}(\mathbf{x})$$

and its Hessian is

$$\nabla_x^2 \phi(\mathbf{x}; \mathbf{v}) + r\nabla\mathbf{h}(\mathbf{x})^{\mathrm{T}}\nabla\mathbf{h}(\mathbf{x}) + r\Sigma_{j=1}^{m} h_j(x)\nabla^2 h_j(x). \tag{5.87}$$

Hence the first order Kuhn–Tucker condition for the original and augmented problems agree at $\bar{\mathbf{x}}$ and $\bar{\mathbf{v}}$ and the Hessian of the augmented problem's Lagrangian is given at $\bar{\mathbf{v}}$ by

$$\mathbf{F}(\bar{\mathbf{x}}) + r\nabla\mathbf{h}(\bar{\mathbf{x}})^{\mathrm{T}}\nabla\mathbf{h}(\bar{\mathbf{x}}). \tag{5.88}$$

Now $\mathbf{F}(\bar{\mathbf{x}})$ is positive definite on the tangent space of the constraints at $\bar{\mathbf{x}}$, i.e. for solutions to the equation $\nabla\mathbf{h}(\bar{\mathbf{x}})\mathbf{x} = \mathbf{0}$. Furthermore, $\nabla\mathbf{h}(\bar{\mathbf{x}})^{\mathrm{T}}\nabla\mathbf{h}(\bar{\mathbf{x}})$ is positive definite for n vectors orthogonal to this tangent space. For large enough r the matrix 5.88 is therefore positive definite everywhere, and local duality theory applies.

In order to evaluate the dual function of the augmented problem at $\mathbf{v}_k$, we must perform an unconstrained minimization of 5.86 with $\mathbf{v} = \mathbf{v}_k$. To see how to steer the sequence of dual function evaluations $\mathbf{v}_k$ rapidly to $\bar{\mathbf{v}}$, note that near to $\bar{\mathbf{x}}$ the third term of 5.87 becomes negligible. Ignoring this term, we have from 5.82 that the Hessian of the dual function of the augmented problem is approximately given by

$$-\nabla\mathbf{h}(\mathbf{x})[\mathbf{F}(\bar{\mathbf{x}}) + r\nabla\mathbf{h}(\mathbf{x})^{\mathrm{T}}\nabla\mathbf{h}(\mathbf{x})]^{-1}\nabla\mathbf{h}(\mathbf{x})^{\mathrm{T}}, \qquad (5.89)$$

which for large r is approximated by $r^{-1}\mathbf{I}$, where $\mathbf{I}$ is the $m \times m$ identity matrix. Moreover, from 5.79 the Jacobian gradient vector of the dual function at $\mathbf{v}_k$ is $\mathbf{h}(\mathbf{x}_k)$, where $\mathbf{x}_k$ is the unconstrained minimum of 5.86 evaluated at $\mathbf{v} = \mathbf{v}_k$. Combining this information, we are led to the approximate Newton–Raphson iteration

$$\mathbf{v}_{k+1} = \mathbf{v}_k + r\mathbf{h}(\mathbf{x}_k)\,. \qquad (5.90)$$

Since the Newton–Raphson iteration converges in a single step when the contours of the function to be optimized are spherical, i.e. its Hessian is a multiple of $\mathbf{I}$, where 5.89 is a good approximation, near to $\bar{\mathbf{x}}$, convergence of the iteration 5.90 can be made arbitrarily fast by taking r large enough. Sequential minimization of 5.86 for a sequence of multiplier vectors $\mathbf{v}_k$ generated by 5.90 has been proposed and tested independently by Hestenes (1969) and Haarhoff and Buys (1970).

We are now in a position to see that Powell's (1969) method introduced in Section 5.6 is equivalent to that of the Hestenes–Haarhoff–Buys algorithm involving dual function evaluations. Taking $r_j = r/2$, $j = 1, \ldots, m$, in 5.44, Powell's method involves the sequential unconstrained minimization of functions F^k given by

$$\begin{aligned} F^k(\mathbf{x}) &= f(\mathbf{x}) + \frac{r}{2}[\mathbf{h}(\mathbf{x}) - \mathbf{s}_k]^{\mathrm{T}}[\mathbf{h}(\mathbf{x}) - \mathbf{s}_k] \\ &= f(\mathbf{x}) + \frac{r}{2}\mathbf{h}(\mathbf{x})^{\mathrm{T}}\mathbf{h}(\mathbf{x}) - r\mathbf{s}_k^{\mathrm{T}}\mathbf{h}(\mathbf{x}) + \frac{r}{2}\mathbf{s}_k^{\mathrm{T}}\mathbf{s}_k \end{aligned} \qquad (5.91)$$

Since $r\mathbf{s}_k^{\mathrm{T}}\mathbf{s}_k/2$ is held constant during the minimization of F^k, the minimum $\mathbf{x}_k$ of F^k also minimizes

$$f(\mathbf{x}) - r\mathbf{s}_k^{\mathrm{T}}\mathbf{h}(\mathbf{x}) + \frac{r}{2}\mathbf{h}(\mathbf{x})^{\mathrm{T}}\mathbf{h}(\mathbf{x})$$

Defining $\mathbf{v}_k = -r\mathbf{s}_k$, Powell's iteration (cf. 4.45) for changing the parameters $\mathbf{s}_k$, namely

$$\mathbf{s}_{k+1} = \mathbf{s}_k - \mathbf{h}(\mathbf{x}_k),$$

can be written as

$$\mathbf{v}_{k+1} = \mathbf{v}_k + r\mathbf{h}(\mathbf{x}_k)$$

which is Hestenes–Haarhoff–Buys iteration 5.90 for adjusting the multipliers $\mathbf{v}_k$ for the augmented problem.

Powell's method can also be considered to involve a sequence of primal function evaluations (cf. Dempster and Rogers, 1973). We saw in Section 5.6 that the minimum $\mathbf{x}_k$ of 5.91 is a solution of the problem of minimizing f subject to the constraint

$$\mathbf{h}(\mathbf{x}) = \mathbf{h}(\mathbf{x}_k).$$

Hence $f(\mathbf{x}_k)$ is an evaluation of the primal function p of the original problem at $\mathbf{h}(\mathbf{x}_k)$. Now for large enough (fixed) r, $\mathbf{h}(\mathbf{x}_k)$ at the unconstrained minimum of 5.91 is obviously arbitrarily close to $\mathbf{s}_k$. Hence $f(\mathbf{x}_k)$ is arbitrarily close to an evaluation of the primal function of the original problem at $\mathbf{s}_k$. Since we know that $p(\mathbf{0}) = f(\bar{\mathbf{x}})$, the obvious method to steer the sequence $\mathbf{s}_k$ of primal function evaluations to $\mathbf{0}$ is to apply a Newton–Raphson iteration to minimize the function with spherical contours given by $\mathbf{s}^{\mathrm{T}}\mathbf{s}$. If p were evaluated at $\mathbf{s}_k$ this iteration would converge to $\mathbf{0}$ from $\mathbf{s}_k$ in a single step of $-\mathbf{s}_k$. Hence by choosing r sufficiently large, the iteration

$$\mathbf{s}_{k+1} = \mathbf{s}_k - \mathbf{h}(\mathbf{x}_k)$$

converges to $\mathbf{0}$, with associated unconstrained minimum $\mathbf{x}_k$ of F^k converging to $\bar{\mathbf{x}}$, arbitrarily fast.

For fixed $\mathbf{s}_k$ the corresponding multiplier vector of the original problem is given in terms of the multiplier vector $\mathbf{v}_k$ of the augmented problem as

$$\tilde{\mathbf{v}}_k = r(\mathbf{h}(\mathbf{x}) - \mathbf{s}_k) = r\mathbf{h}(\mathbf{x}) + \mathbf{v}_k$$

and the Hessian of F^k is given by

$$\nabla^2 F^k(\mathbf{x}) = \nabla_x^2 \phi(\mathbf{x}, \tilde{\mathbf{v}}_k) + r\nabla \mathbf{h}(\mathbf{x})^T \nabla \mathbf{h}(\mathbf{x})$$
$$= \nabla_x^2 \phi(\mathbf{x}, \mathbf{v}_k) + r\nabla \mathbf{h}(\mathbf{x})^T \nabla \mathbf{h}(\mathbf{x}) + r\Sigma_{j=1}^m h_j(\mathbf{x})\nabla^2 h_j(\mathbf{x}).$$

We have seen that $\nabla^2 F^k$ is positive definite near the constrained minimum $\bar{\mathbf{x}}$ for large enough r. The practical convergence of the Hestenes–Haarhoff–Buys–Powell method far from $\bar{\mathbf{x}}$ depends only on the requirement of positive definiteness of $\nabla^2 F^k$ in order to ensure rapid convergence of the sequential minimizations. When it is used with a quasi-Newton algorithm which maintains a positive definite approximation to $(\nabla^2 F^k)^{-1}$, it can therefore be expected to be extremely efficient. Note that $\nabla^2 F^k$ will become ill-conditioned (in the sense of possessing uniformly large eigenvalues) near to the constrained optimum only if the problem is irregular, i.e. $\mathbf{v}_k = -\nabla p(\mathbf{s}_k)$ tends in all co-ordinates to infinity. (In practice, of course, Powell's method adjusts r and $\mathbf{s}$ to improve convergence in this situation, see Section 5.6.)

There are alternative strategies in extending the primal or dual function evaluation algorithm to handle the general problem involving inequality constraints. A first is to combine the equation constrained version with the active set strategy as in Osborne and Ryan's hybrid algorithm for Powell's primal method. Another strategy is to handle an inequality constraint $g_i(\mathbf{x}) \leqslant 0$ as an equation $g_i(\mathbf{x}) + y_i^4 = 0$ through the use of a slack variable y_i. The resulting F^k (alternatively, the augmented Lagrangian) is easily seen to be twice continuously differentiable under similar assumptions on the problem functions and to be stationary in the slack variables y_i at the constrained optimum. The difficulty with this approach is that even though they are easy to evaluate, viz.

$$y_i = |g_i(\mathbf{x})|^{\frac{1}{4}} \text{ when } g_i(\mathbf{x}) < 0 \text{ and } y_i = 0 \text{ when } g_i(\mathbf{x}) \geqslant 0,$$

m new variables are introduced.

As a result, a number of workers have been led to consider a generalized Lagrangian of the form

$$\psi(\mathbf{x}, \mathbf{y}, \mathbf{z}) = f(\mathbf{x}) + \Sigma_{i=1}^l G(r, g_i(\mathbf{x}), y_i) + \Sigma_{j=1}^m H(r, h_j(\mathbf{x}), z_j)$$

in which the dual variable vectors $\mathbf{y}$ and $\mathbf{z}$ enter non-linearly. Following early work by Arrow and Solow (1958), these are required to have some or all of the properties of the ordinary Lagrangian in which the multiplier vectors $\mathbf{u}$ and $\mathbf{v}$ enter linearly. Rockafellar (1973) has given a specific generalized Lagrangian form whose *global* saddlepoints correspond to *global* solutions of the original problem. However it is only once globally differentiable in the variables $\mathbf{y}$ and $\mathbf{z}$, and is therefore unsuitable for use with quasi-Newton algorithms. Mangasarian (1973) defines a class of generalized Lagrangian functions which are stationary in all variables at a constrained minimum $\mathbf{x}$, are globally twice continuously differentiable when the problem functions are, and whose Hessian $\nabla_x^2 \psi(x, \bar{y}, \bar{z})$, evaluated at the optimal dual variable vectors $\bar{\mathbf{y}}$, $\bar{\mathbf{z}}$, is positive definite in $\mathbf{x}$ at $\bar{\mathbf{x}}$ for large enough r. The optimal dual variables $\bar{\mathbf{y}}$ and $\bar{\mathbf{z}}$ are related to the optimal Lagrange multipliers $\bar{\mathbf{u}}$ and $\bar{\mathbf{v}}$ by the equations

$$\bar{u}_i = \frac{\partial}{\partial g_i} G(r, g_i(\bar{\mathbf{x}}), \bar{\mathbf{y}}_i) \qquad i = 1, \ldots, l$$

$$\bar{v}_j = \frac{\partial}{\partial h_j} H(r, h_j(\bar{\mathbf{x}}), \bar{\mathbf{z}}_j) \qquad j = 1, \ldots, m$$

Moreover, the original problem has a dual problem analogous to 5.85 of the form

$$\text{maximize } \psi(\mathbf{x}, \mathbf{y}, \mathbf{z})$$

subject to

$$\nabla_x \psi(\mathbf{x}, \mathbf{y}, \mathbf{z}) = \mathbf{0}.$$

Mangasarian proposes a sequential unconstrained minimization procedure in which ψ is minimized at $\mathbf{x}_k$ for fixed values of the dual variables $\mathbf{y}_k$, $\mathbf{z}_k$ and the latter are updated by the iterations

$$y_i^{k+1} = \begin{cases} y_i^k + \beta \partial \psi(\mathbf{x}_k, \mathbf{y}_k, \mathbf{z}_k)/\partial y_i & \text{if } g_i(x_k) \geqslant 0 \\ 0 & \text{if } g_i(x_k) < 0 \end{cases}$$

$$z_j^{k+1} = z_j^k + \beta \partial \psi(\mathbf{x}_k, \mathbf{y}_k, \mathbf{z}_k)/\partial z_j \qquad j = 1, \ldots, m.$$

He demonstrates linear convergence of this process to an

isolated local constrained minimum $\bar{\mathbf{x}}$ for some $\beta > 0$. The choice of β depends on the condition number of $\nabla_x^2 \psi$ at the constrained optimum; for large or small values of r this will be large, forcing β to be small and convergence to be slow. (Mangasarian also suggests applying Newton's method to ψ as a function of $\mathbf{x}$, $\mathbf{y}$ and $\mathbf{z}$, which will converge superlinearly or quadratically according as the problem functions are twice or three times continuously differentiable.) Although largely untested, generalized Lagrangian methods appear promising.

Another approach to extending the primal and dual function evaluation algorithms to inequality constraints is to apply dual function evaluation, but to restrict the iteration 5.90 to be non-negative. Sayama *et al* (1973) propose a twice-continuously differentiable generalized Lagrangian involving a G function of the form

$$G(r, g_i(\mathbf{x}), y_i) = \begin{cases} rg_i^2(\mathbf{x}) + y_i g(\mathbf{x}) & \text{if } g_i(\mathbf{x}) \geqslant 0 \\ \dfrac{y_i^2 g_i(\mathbf{x})}{y_i - rg_i(\mathbf{x})} & \text{if } g_i(\mathbf{x}) < 0, \end{cases}$$

in which the dual variables y for the inequality constraints are maintained non-negative by the iteration

$$y_i^{k+1} = \begin{cases} y_i^k + rg_i(\mathbf{x}_k) & \text{if } g_i(\mathbf{x}_k) \geqslant 0 \\ \dfrac{(y_i^k)^3}{[y_i^k - rg_i(\mathbf{x}_k)]^2} & \text{if } g_i(\mathbf{x}_k) < 0. \end{cases}$$

Theoretical convergence to a constrained minimum is established, ill-conditioning is avoided and the numerical results appear promising. Primal function evaluation could be approximated by minimizing Powell's F^k subject to linearizations of the inequality constraints using a suitable linearly constrained routine. This strategy would however result in an algorithm belonging to a class which we will now describe briefly.

Sequential linearly constrained techniques

Two very similar Lagrangian techniques have recently been proposed which solve a sequence of linearly constrained problems using the second order methods of Section 5.7. The first, due to Robinson (1972), is for the general constrained problem. The constraints

$$\mathbf{g}(\mathbf{x}) \leqslant \mathbf{0} \quad \text{and} \quad \mathbf{h}(\mathbf{x}) = \mathbf{0}$$

are linearized about the current point $\mathbf{x}_k$ as

$$\mathbf{g}(\mathbf{x}_k) + \nabla\mathbf{g}(\mathbf{x}_k)(\mathbf{x} - \mathbf{x}_k) \leqslant \mathbf{0} \text{ and } \mathbf{h}(\mathbf{x}_k) + \nabla\mathbf{h}(\mathbf{x}_k)(\mathbf{x} - \mathbf{x}_k) = \mathbf{0}. \tag{5.92}$$

To find $\mathbf{x}_{k+1}$ given the current estimate of the multiplier vectors $\mathbf{u}_k$, $\mathbf{v}_k$, the kth iteration minimizes the Lagrangian given by

$$\phi(\mathbf{x}; \mathbf{u}_k, \mathbf{v}_k) = f(\mathbf{x}) + \mathbf{u}_k^{\mathrm{T}}\mathbf{g}(\mathbf{x}) + \mathbf{v}_k^{\mathrm{T}}\mathbf{h}(\mathbf{x}) \tag{5.93}$$

subject to the linear constraints 5.92 using the algorithm of Ritter (1972). The multiplier vectors $\mathbf{u}_{k+1}$, $\mathbf{v}_{k+1}$ are taken to be the optimal multipliers for the linearized constraints. A similar method in which 5.93 was replaced by its quadratic approximation at $\mathbf{x}_k$ was proposed earlier by Wilson (1963). The Kuhn–Tucker conditions for the original problem and its linearization are identical at the constrained optimum $\bar{\mathbf{x}}$. At $\mathbf{x}_k$ they will agree up to second order terms of the form $|\mathbf{x} - \mathbf{x}_k|^2$, $|\mathbf{u} - \mathbf{u}_k|^2$ and $|\mathbf{v} - \mathbf{v}_k|^2$, providing, of course, that the problem functions are twice continuously differentiable.

Rosen and Kreuser's (1972) method for the inequality constrained problem is identical to Robinson's except that Goldfarb's (1969) linearly constrained algorithm is used and the optimal multipliers for the kth linearized problem are not computed. Instead, the solution of the Kuhn–Tucker condition

$$\nabla f(\mathbf{x}_{k+1}) + \mathbf{u}_{k+1}^{\mathrm{T}}\nabla\mathbf{g}(\mathbf{x}_{k+1}) = \mathbf{0}$$

is used to give the co-ordinates of $\mathbf{u}_{k+1}$ corresponding to the s violated constraints at $\mathbf{x}_{k+1}$ as

$$\mathbf{w}_{k+1} = -\mathbf{K}^{+}\mathbf{g} = -(\mathbf{K}\mathbf{K}^{\mathrm{T}})^{-1}\mathbf{K}\nabla f(\mathbf{x}_{k+1})^{\mathrm{T}}, \tag{5.94}$$

where **K** is the $s \times m$ matrix of active constraint gradients, $\mathbf{K}^+$ denotes its generalized inverse and **g** is the Jacobian gradient vector of f. Rosen and Kreuser's method was proposed for convex problems, but even for these, the second order expression for the current multipliers provided by Ritter's algorithm in Robinson's method are superior.

Quadratic convergence in major iterations has been established for both methods and both have been tested on some standard test problems. These limited results indicate that Robinson's code is marginally superior to Abadie's GRG 69 code.

Unconstrained methods

Fletcher (1970a) has proposed a class of Lagrangian methods which solve a constrained problem in a single unconstrained minimization. The basic idea involves versions of the Lagrangian function in which the multipliers are approximated by functions of **x** in such a way as to make the resulting function a convex function of **x** with a global minimum at a constrained minimum of the original problem. Thus only a single minimization is required.

Consider the active equation constraints $\mathbf{k}(\mathbf{x}) = \mathbf{0}$, and as usual denote by $\mathbf{K}^+$ the $n \times s$ generalized inverse $\mathbf{K}^T(\mathbf{K}\mathbf{K}^T)^{-1}$ of the $s \times n$ matrix **K** of constraint function gradients $\nabla \mathbf{k}_r$. Although other such *exact* penalty functions are possible, the function given by

$$\psi(\mathbf{x}) = f(\mathbf{x}) + \mathbf{u}^T k(\mathbf{x})$$

with

$$\mathbf{u} = \mathbf{K}^{+T}[-\mathbf{g} + r\mathbf{K}^+\mathbf{k}(\mathbf{x})]$$

has been considered most extensively by Fletcher. An alternative interpretation of ψ, made clear by substituting 5.96 in 5.95, is that it is the ordinary Lagrangian of the problem $\phi = f + \mathbf{w}^T\mathbf{k}$ augmented by the positive definite quadratic form $r\mathbf{k}^T\mathbf{K}^{+T}\mathbf{K}^+\mathbf{k}$ with **w** given by expression 5.56. Fletcher has shown that if the solution of the problem satisfies the second order conditions 5.14 to 5.17 the Hessian matrix $\nabla^2\psi$ of ψ is positive definite for large enough r. The required magnitude of r can be determined in terms of the elements of $\nabla^2 \psi$.

Second order constrained methods are used to minimize ψ for equality constrained problems. The resulting algorithms are designed to use only second derivative information in spite of the third derivatives entering into $\nabla^2 \psi$. They are second order constrained methods, with super-linear (usually quadratic) rates of convergence to a point satisfying the second order conditions 5.14 to 5.17 (Fletcher, 1972b). Moreover, they solve a quadratic programme, i.e. a quadratic problem with linear constraints, in a finite number of steps, see Fletcher (1970a, 1971). Fletcher (1972c) has shown that for inequality constrained problems the value of $\mathbf{u}$ given by 5.96 at a constrained minimum is the solution of a simple quadratic programming problem. He proposes an algorithm which solves such a problem at each iteration using an updated set of currently binding constraints.

In limited tests on equality constrained problems, Fletcher and Lill (1970) report performance comparable to sequential unconstrained minimization techniques. Fletcher (1972c) reports encouraging results in some tests of his inequality constrained version, but notes that most of the iterations were used to locate the constraints which bind at the constrained minimum (see also Lill, 1972). This could be taken to support the view that one of the factors contributing to the success of Abadie's GRG code is its treatment of all inequality constraints as equations through the use of slack variables.

5.9 General considerations

Colville (1970) reports the compilation of tests on a set of nine standard non-linear problems using some thirty computer codes for non-linear optimization. Although the problems are mostly constrained, some purely unconstrained algorithms were tested. The non-linear functions involved ranged from quadratics to the output of a self-contained simulation model (see Exercise 10). The time to run a multiple matrix inversion test program was used to standardize all run times, but no attention was paid to standardizing accuracy as this was considered too difficult when each participant ran the problems on his own computer. Colville found that the number of function evaluations proved a totally unreliable guide to run time. Since Colville's study, new methods and their codes have been tested on some or all of Colville's problems. Robinson's (1972) linearly constrained Lagrangian code is currently fastest for three of them and comparable to GRG 69 for a fourth. The relative efficiencies of the algorithm classes for the constrained problems were in the order we have studied them, with the GRG 69 code of Abadie clearly the fastest overall. Abadie (1969) reports an accuracy of 10^{-6} to 10^{-8} for this code.

A number of Colville's caveats in interpreting such experiments should be pointed out. Run times of the standard test program on identical hardware varied up to 10 per cent due to the varying compiling efficiency of individual computer systems. (In this context it should be noted that run times of different codes on a given problem varied by factors of up to 160.) More important, Colville found that different codings of the same basic algorithm varied by large factors. This shows the importance of sophistication in the numerical methods embodied in, for example, matrix inversion. More generally, it emphasizes the importance of efficient organization of the calculations. Finally, results on a set of small problems will be distorted by the data manipulation required by a production code designed for the routine solution of large problems when these compete with experimental codes set up for small problems.

Theoretical considerations for evaluating the efficiency of constrained optimization routines are available and McCormick (1971) contains a discussion of some of these. A natural requirement is that feasibility with respect to inequality constraints be maintained at each iteration of the process. With the exception of exterior point penalty function methods and sequential linearly constrained Lagrangian techniques, the algorithms we have outlined all possess this property. A second desirable property of an algorithm is that it contains no parameters that must be set arbitrarily, either by the computer code or the user. Many unconstrained methods are parameter-free because they involve a linear search or a step size of one (as in the newer quasi-Newton methods). However, with the exception of some of the Lagrangian techniques, all the constrained methods above require arbitrary parameter choices in addition to the usual specification of tolerances, rather than adjusting these choices in the light of current problem information at each iteration. A related criterion proposed by Wolfe (1969a) concerns the performance of a non-linear constrained technique, without modification to take account of special structure, when all the problem functions are in fact linear. In this situation, an efficient routine could be expected to compete with the simplex method for linear programming by finding a solution in a finite number of iterations. Fletcher (1970a) suggests requiring a similar property for problems with a quadratic objective function and linear constraints with regard to efficient algorithms for quadratic programming. Since they must move interior to linear constraints, direct search and small step gradient methods do not have these properties; sequential unconstrained techniques clearly do not either. Of the methods examined above, only large step gradient methods and the sequential linearly constrained methods, which incorporate large step gradient methods, meet these criteria.

The last three criteria of the previous paragraph are special aspects of the general proposition that an efficient procedure should use current problem information from one iteration to the next to reduce the order of arithmetic operations required. Small step gradient methods and the direct search methods popular in engineering design applications fail to do this. On the other hand, extrapolation techniques used in practice with sequential unconstrained and large step gradient and Lagrangian methods have been developed precisely to meet this point. In operational codes for *un*constrained minimization, a widespread practice is to combine methods in series so as to efficiently exploit current problem information to accelerate convergence to a minimum. A typical sequence begins with a first order gradient method to locate the area of a local minimum, followed by a search technique using conjugate directions until few new directions are produced, when a neighbourhood of a minimum is identified. At that point a few iterations of a second order gradient method usually suffice to approximate the minimum accurately. Further, (if it is required) linear search is almost always effected by quadratic interpolation in phases one and three; in the former because a rough estimate will suffice, in the latter because in a small neighbourhood of the minimum it is almost exact. Little use of sequencing of appropriate algorithms for constrained minimization seems to have been made at present, although the idea could easily be implemented with methods making use of unconstrained optimization techniques at each iteration.

Published proofs of theoretical convergence are not currently available for all the algorithms for constrained minimization described above. However, under our regularity assumption, and with the proper numerical precautions against zig-zagging it is possible to prove that they converge to a point satisfying the Kuhn–Tucker first order necessary conditions 5.7 to 5.9. Zangwill (1969) has given a general theory for effecting this. Unfortunately, many of the current constrained methods come unstuck when the particular regularity assumption we have used fails to hold, even though the satisfaction of the Kuhn–Tucker conditions by local constrained minima remains valid more generally, see Dempster (1973), Chapter 7. Current research has begun the attack on efficient procedures for handling such *degenerate* situations, but with the exception of sequential unconstrained techniques little or no work concerns algorithms which will locate *irregular* solutions where weaker conditions than the Kuhn–Tucker conditions are satisfied.

For problems in which all functions are convex, we have seen that solutions of the Kuhn–Tucker conditions are global minima. For this reason, with the exception of the work of Fiacco and McCormick (1968) and Fletcher (1969–72), little attention has been paid until recently to convergence of unconstrained methods to points satisfying the second order conditions 5.11 to 5.13 or 5.14 to 5.17. Second order constrained methods, i.e. those possessing this property, should be amenable to an analysis of the rate of convergence corresponding to those available for second order methods for unconstrained optimization. McCormick (1971) has suggested that close to a point satisfying the second order conditions, i.e. once the binding inequality constraints have been identified, any efficient algorithm should effect a step similar to the classical Newton's method applied to solve the Kuhn–Tucker conditions 5.7 to 5.9. Few such analyses of the efficient large step gradient and Lagrangian methods have so far been published, although Fletcher's algorithms (1971, 1972) have been designed to have this property. A simple requirement for accelerating convergence early in the computational process, violated only by sequential unconstrained techniques, is the stability requirement that the value E of the objective function f should decrease at each iteration.

5.10 Conclusion

The optimization methods presented in this book have been illustrated by examples and exercises which involve simple and relatively well behaved mathematical functions. This is far removed from the use to which optimization methods are put in practice. The only way to appreciate the problems involved in practical situations is to apply the techniques available to a number of examples drawn from a familiar field of application. The behaviour of the various optimization methods should then yield some insight into their suitability for future typical problems in the field.

Exercises 5

1 Repeat question 3 of the exercises on Chapter 1.

2 Use library subroutines, or write a code for some or all of the principal methods of this chapter, to find the minimum of

$$E = 100(x_2 - x_1^2)^2 + (1 - x_1)^2 + 90(x_4 - x_3^2)^2 + (1 - x_3)^2 + 10.1(x_2 - 1)^2 + (x_4 - 1)^2 + 19.8(x_2 - 1)(x_4 - 1)$$

subject to the constraints $-10 \leqslant x_i \leqslant 10, i = 1, \ldots, 4$ (Colville (1968), # 6, Wood). Begin at $(-3, -1, -3, -3)$.

3 Repeat Exercise 2 with the following problem due to Rosenbrock (1960).
What is the maximum volume of a parcel to be sent by post? The maximum length is 42 inches and the maximum girth plus length is 72 inches.

4 Repeat with objective function given by

$$E = x_1^2 + 4x_2^2$$

and a linear constraint $x_1 + 2x_2 = 1$ (Fletcher, 1969). Begin at $(1, 1)$.

5 Repeat with objective function given by

$$E = \Sigma_{i=1}^{16} \Sigma_{j=1}^{16} a_{ij}(x_i^2 + x_i + 1.0)(x_j^2 + x_j + 1.0)$$

subject to the constraints $0 \leqslant x_i \leqslant 5, i = 1, \ldots, 16$, and

$$0.22x_1 + 0.20x_2 + 0.19x_3 + 0.25x_4 + 0.15x_5 + 0.11x_6 + 0.12x_7 + 0.13x_8 + x_9 = 2.5$$

$$-1.46x_1 - 1.3x_3 + 1.82x_4 - 1.15x_5 + 0.8x_7 + x_{10} = 1.1$$

$$1.29x_1 - 0.89x_2 - 1.16x_5 - 0.96x_6 - 0.49x_8 + x_{11} = -3.1$$

$$-1.10x_1 - 1.06x_2 + 0.95x_3 - 0.54x_4 - 1.78x_6 - 0.41x_7 + x_{12} = -3.5$$

$$-1.43x_4 + 1.51x_5 + 0.59x_6 - 0.33x_7 - 0.43x_8 + x_{13} = 1.3$$

$$-1.72x_2 - 0.33x_3 + 1.62x_5 + 1.24x_6 + 0.21x_7 - 0.26x_8 + x_{14} = 2.1$$

$$1.12x_1 + 0.31x_4 + 1.12x_7 - 0.36x_9 + x_{15} = 2.3$$

$$0.45x_2 + 0.26x_3 - 1.10x_4 + 0.58x_5 + 1.03x_7 + 0.1x_8 + x_{16} = -1.5$$

where twice the coefficients a_{ij} in the objective function are given below (Colville #7, Sauthier). Begin at **0**.

i \ j	1	2	3	4	5	6	7	8	9	10	11	12	13	14	15	16
1	2.0			1.0			1.0	1.0								1.0
2		2.0	1.0				1.0			1.0						
3		1.0	2.0				1.0		1.0	1.0				1.0		
4	1.0			2.0			1.0				1.0				1.0	
5					2.0	1.0				1.0		1.0				1.0
6					1.0	2.0		1.0							1.0	
7	1.0	1.0	1.0	1.0			2.0				1.0		1.0			
8	1.0					1.0		2.0		1.0					1.0	
9			1.0						2.0			1.0				1.0
10		1.0	1.0		1.0			1.0		2.0				1.0		
11				1.0			1.0				2.0		1.0			
12					1.0				1.0			2.0		1.0		
13							1.0				1.0		2.0	1.0		
14			1.0							1.0		1.0	1.0	2.0		
15				1.0				1.0							2.0	
16	1.0				1.0				1.0							2.0

6 Repeat with objective function given by

$$E = 5.3578547x_3^2 + 0.8356891x_1x_5 + 37.293239x_1 - 40792.141$$

and constraints

$$0 \leqslant 85.334407 + 0.0056858x_2x_5 + 0.0006262x_1x_4 - 0.0022053x_3x_5 \leqslant 92$$

$$90 \leqslant 80.51249 + 0.0071317x_2x_5 + 0.0029955x_1x_2 + 0.0021813x_3^2 \leqslant 110$$

$$20 \leqslant 9.300961 + 0.0047026x_3x_5 + 0.0012547x_1x_3 + 0.0019085x_3x_4 \leqslant 25$$

$78 \leqslant x_1 \leqslant 102$ $\quad$ $33 \leqslant x_2 \leqslant 45$

$27 \leqslant x_3 \leqslant 45$ $\quad$ $27 \leqslant x_4 \leqslant 45$

$27 \leqslant x_5 \leqslant 45$

(Colville #3, Grace). Begin at

$$(78.62,\ 33.44,\ 31.07,\ 44.18,\ 35.32)$$

and

$$(78.0,\ 33.0,\ 27.0,\ 27.0,\ 27.0).$$

7 Repeat with objective function given by

$$E = \Sigma_{j=1}^{5} e_j y_j + \Sigma_{j=1}^{5} \Sigma_{i=1}^{5} c_{ij} y_i y_j + \Sigma_{j=1}^{5} d_j y_j^3$$

and constraints

$$\Sigma_{j=1}^{5} a_{ij} y_j \geqslant b_i \qquad i = 1, \ldots, 10$$

$$y_j \geqslant 0 \qquad j = 1, \ldots, 5$$

where the coefficients are given in the table below (Colville #1).
Begin at (0,0,0,0,1).

	i	j 1	2	3	4	5	
e_j		−15	−27	−36	−18	−12	
c_{ij}	1	30	−20	−10	32	−10	
	2	−20	39	−6	−31	32	
	3	−10	−6	10	−6	−10	
	4	32	−31	−6	39	−20	
	5	−10	32	−10	−20	30	
d_j		4	8	10	6	2	b_i
a_{ij}	1	−16	2	0	1	0	−40
	2	0	−2	0	0.4	2	−2
	3	−3.5	0	2	0	0	−.25
	4	0	−2	0	−4	−1	−4
	5	0	−9	−2	1	−2.8	−4
	6	2	0	−4	0	0	−1
	7	−1	−1	−1	−1	−1	−40
	8	−1	−2	−3	−2	−1	−60
	9	1	2	3	4	5	5
	10	1	1	1	1	1	1

8 Repeat with the dual of the previous problem (Colville #2) whose objective function is given by

$$E = \Sigma_{i=1}^{10} b_i x_i + \Sigma_{i=1}^{5} \Sigma_{j=i}^{5} c_{ij} y_i y_j + 2\, \Sigma_{j=1}^{5} d_j y_j^3$$

and whose constraints are

$$\Sigma_{i=1}^{10} a_{ij} x_i \leqslant e_j + 2\,\Sigma_{i=1}^{5} c_{ij} y_i + 3 d_j y_j^2,\ y_j \geqslant 0 \quad j = 1, \ldots, 5$$

$$x_i \geqslant 0 \qquad i = 1, \ldots, 10.$$

Begin at

$$x_i = 0.0001,\ i \neq 7, \qquad i = 1, \ldots, 10$$

$$x_7 = 60$$

$$y_j = 0.0001, \qquad j = 1, \ldots, 5$$

and

$$y_1 = y_2 = y_3 = y_4 = 0$$

$$y_5 = 1$$

$$x_i = b_i, \qquad i = 1, \ldots, 10.$$

9 Repeat with objective function given by

$$E = x_1 + x_2$$

and constraints $-20 \leqslant x_1, x_2 \leqslant 20$ and

$$\exp\,[(-x_1^2 - x_2^2 - 4)/4] - \exp\,[(-x_1^2 - x_2^2)/2] = 0$$

(Crabill, Evans and Gould, 1971). Begin at (2, 0).

10 Repeat with objective function given by

$$E = 0.063 y_2 y_5 - 5.04 x_1 - 3.36 y_3 - 0.035 x_2 - 10.0 x_3$$

and constraints

$$0 \leqslant x_1 \leqslant 2000.0 \qquad 0 \leqslant y_2 \leqslant 5000.0$$
$$0 \leqslant x_2 \leqslant 16{,}000.0 \qquad 0 \leqslant y_3 \leqslant 2000.0$$
$$0 \leqslant x_3 \leqslant 120.0 \qquad 85.0 \leqslant y_4 \leqslant 93.0$$
$$90.0 \leqslant y_5 \leqslant 95.0$$
$$3.0 \leqslant y_6 \leqslant 12.0$$
$$0.01 \leqslant y_7 \leqslant 4.0$$
$$145.0 \leqslant y_8 \leqslant 162.0$$

where the dependent x variables are calculated from the independent y variables according to the FORTRAN code given below (Colville #9).
Begin at (1745.0, 12,000.0, 110.0).

FORTRAN Description of Calculation of Y(2) → Y(8)

```
       Y(2) = 1.6*X(1)
 10    Y(3) = 1.22*Y(2) – X(1)
       Y(6) = (X(2) + Y(3))/X(1)
       Y2CALC = X(1)*(112.0 + 13.167*Y(6)
                – 0.6667*Y(6)**2)/100.0
       IF(ABS(Y2CALC – Y(2)) – 0.001)30,30,20
 20    Y(2) = Y2CALC
       GO TO 10
 30    CONTINUE
       Y(4) = 93.0
 100   Y(5) = 86.35 + 1.098*Y(6) – 0.038*Y(6)**2
             + 0.325*(Y(4) – 89.0)
       Y(8) = – 133.0 + 3.0*Y(5)
       Y(7) = 35.82 – 0.222*Y(8)
       Y(4)CALC = 98000.0*X(3)/(Y(2)*Y(7) + X(3)*1000.0)
       IF(ABS(Y4CALC – Y(4)) – 0.0001)300,300,200
 200   Y(4) = Y4CALC
       GO TO 100
 300   CONTINUE
```

References

Abbreviations

Compt. Rend. Acad. Sci. (Paris)	*Comptes Rendus d'Academie des Sciences (Paris)*
Computer J.	*Computer Journal*
J. ACM.	*Journal of the Association for Computing Machinery*
J. Inst. Math. Applns.	*Journal of the Institute for Mathematics and Its Applications*
J. Math. Anal. Applns.	*Journal of Mathematical Analysis and Its Applications*
J. Opt. Theory Applns.	*Journal of Optimization Theory and Its Applications*
J. Res. Natl. Bureau Standards	*Journal of Research of the (U.S.) National Bureau of Standards*
J. SIAM.	*Journal of the Society for Industrial and Applied Mathematics*
Math. Computing	*Mathematics of Computing*
Math. Prog.	*Mathematical Programming*
Managmt. Sci.	*Management Science*
Opns. Res.	*Operations Research*
Quart. J. Appl. Math.	*Quarterly Journal of Applied Mathematics*
SIAM. J. Appl. Math.	*Society for Industrial and Applied Mathematics Journal of Applied Mathematics*
SIAM Rev.	*Society for Industrial and Applied Mathematics Review*

[1] Abadie, J. (1967) ed., *Nonlinear Programming*. North Holland: Amsterdam.

[2] Abadie, J. (1970) ed., *Integer and Nonlinear Programming*. North Holland: Amsterdam.

[3] Abadie, J. (1970a) Application of the GRG algorithm to optimal control problems, in [2], 191–211.

[4] Abadie, J. & Carpentier, J. (1965) Généralization de la méthode du gradient réduit de Wolfe au cas des contraintes non-linéares, EDF note HR6678, reprinted as 'Generalization of the Wolfe reduced gradient method to the case of nonlinear constraints' in [34], 37–47.

[5] Abadie, J. & Guigou, J. (1970) Numerical experiments with the GRG method, Appendix III in [2], 529–536.

[6] Anderssen, R.S. (1972) Global optimization, in [7], 26–48.

[7] Anderssen, R.S., Jennings, L.S. & Ryan D.M. (1972) *Optimization*. U. of Queensland Press: St. Lucia, Queensland.

[8] Apostal, T.M. (1957) *Mathematical Analysis, A Modern Approach to Advanced Calculus*. Addison–Wesley: Reading, Mass.

[9] Beale, E.M.L. (1972) A derivation of conjugate gradients, in [72], 39–43.

[10] Beltrami, E.J. (1969) A constructive proof of the Kuhn–Tucker multiplier rule. *J. Math. Anal. Applns*, **26**, 297–306.

[11] Beltrami, E.J. (1970) *An Algorithmic Approach to Nonlinear Analysis and Optimization*. Academic: New York.

[12] Birkhoff, G.D. & Maclane, S. (1953) *A Survey of Modern Algebra*. Macmillan: New York.

[13] Box, M.J. (1966) A comparison of several current optimization methods, and the use of transformations in constrained problems. *Computer J*, **8**, 67–77.

[14] Box, M.J., Davies D. & Swann, W.H. (1969) *Non-linear Optimization Techniques*. Oliver & Boyd: Edinburgh.

[15] Brown, K.M. & Dennis, Jr., J.E. (1970) Derivative-free analogues of the Levenberg–Marquardt and Gauss algorithms for non-linear least-squares approximation. IBM Philadelphia Sci. Center Technical Report No. 320–2994.

[16] Broyden, C.G. (1965) A class of methods for solving nonlinear equations. *Math. Computing*, **19**, 577–584.

[17] Broyden, C.G. (1967) Quasi–Newton methods and their application to function minimization. *Math. Computing*, **21**, 368–381.

[18] Broyden, C.G. (1970) The convergence of a class of double-rank minimization algorithms, Parts I & II. *J. Inst. Maths. Applns.*, **6**, 66–90 & 222–231.

[19] Carroll, C.W. (1961) The created response surface technique

for optimizing nonlinear restrained systems. *Opns. Res.*, **9**, 169–185.

[20] Coggins, G.F. (1964) Univariate search methods. ICI Central Instrument Res. Lab. Research Note No. 64/11.

[21] Colville, A.R. (1968) A comparative study on nonlinear programming codes. IBM New York Sci. Center Technical Report No. 320–2949.

[22] Colville, A.R. (1970) A comparative study of nonlinear programming codes, in [66], 487–501.

[23] Crabill, T.B., Evans, J.P. & Gould, F.J. (1971) An example of an ill-conditioned NLP problem. *Math. Prog.*, **1**, 113–116.

[24] Curry, H. (1944) The method of steepest descent for nonlinear minimization problems. *Quart. J. Appl. Math.*, **2**, 258–261.

[25] Davidon, W.C. (1959) Variable metric method for minimization. AEC (U.S.) Research and Development Report No. ANL 5990.

[26] Davidon, W.C. (1969) Variance algorithms for minimization, in [34], 13–20.

[27] Dempster, M.A.H. (1975) *Elements of Optimization.* Chapman & Hall: London.

[28] Dempster, M.A.H. & Rogers, B. (1973) Constrained nonlinear optimization using unconstrained methods. Mathematical Institute, Oxford (mimeo).

[29] Dixon, L.C.W. (1972) Quasi–Newton algorithms generate identical points. *Math. Prog.*, **2**, 383–387.

[30] Dixon, L.C.W. (1972a) The choice of step length, a crucial factor in the performance of variable metric algorithms, in [72], 149–170.

[31] Evans, J.P. & Gould, F.J. (1972) On using equality-constraint algorithms for inequality constrained problems. *Math. Prog.*, **2**, 324–329.

[32] Fiacco, A.V. & McCormick, G.P. (1968) *Nonlinear Programming: Sequential Unconstrained Minimization Techniques.* Wiley: New York.

[33] Fletcher, R. (1965) Function minimization without evaluating derivatives – a review. *Computer J.*, **8**, 33–41.

[34] Fletcher, R. (1969) ed., *Optimization.* Academic: New York.

[35] Fletcher, R. (1969a) A review of methods for unconstrained optimization, in [34], 1–12.

[36] Fletcher, R. (1970) A new approach to variable metric algorithms. *Computer J.*, **13**, 317–322.

[37] Fletcher, R. (1970a) A class of methods for nonlinear programming with termination and convergence properties, in [2], 157–175.

[38] Fletcher, R. (1970b) An efficient, globally convergent, algorithm

for unconstrained and linearly constrained problems. AERE (U.K.) Harwell Technical Report No. TP 431.

[39] Fletcher, R. (1971) A modified Marquardt subroutine for non-linear squares. AERE (U.K.) Harwell Technical Report No. R6799.

[40] Fletcher, R. (1971a) A general quadratic programming algorithm. *J. Inst. Maths. Applns.*, 7, 76–91.

[41] Fletcher, R. (1972) An algorithm for solving linearly constrained optimization problems. *Math. Prog.*, **2**, 133–165.

[42] Fletcher, R. (1972a) Minimizing general functions subject to linear constraints, in [72], 279–296.

[43] Fletcher, R. (1972b) A class of methods for non-linear programming: III. Rates of convergence, in [72], 371–382.

[44] Fletcher, R. (1972c) An exact penalty function for nonlinear programming with inequalities. A.E.R.E. (U.K.) Harwell Technical Report No. TP 478.

[45] Fletcher, R. & Lill, S.A. (1970) A class of methods for nonlinear programming: II. Computational experience, in [119], 67–92.

[46] Fletcher, R. & Powell, M.J.D. (1963) A rapidly convergent descent method for minimization. *Computer J.*, **6**, 163–168.

[47] Fletcher, R. & Reeves, C.M. (1964) Function minimization by conjugate gradients. *Computer J.*, **7**, 149–154.

[48] Fox, L. & Mayers, D.F. (1968) *Computing Methods for Scientists and Engineers*. Clarendon Press: Oxford.

[49] Fox, L. & Parker, J.B. (1968) *Chebyshev Polynomials in Numerical Analysis*. Oxford U. Press: London.

[50] Gill, P.E. & Murray, W. (1971) Quasi–Newton methods for unconstrained optimization. NPL (U.K.) Mathematics Report No. 97.

[51] Gillespie, R.P. (1954) *Partial Differentiation*. Oliver & Boyd: London.

[52] Goldfarb, D. (1969) Extension of Davidon's variable metric algorithm to maximization under linear inequality and equality constraints. *SIAM J. Appl. Maths.*, **17**, 739–764.

[53] Goldfeld, S.M. & Quandt, R.E. (1972) *Nonlinear Methods in Econometrics*. North Holland: Amsterdam.

[54] Goldfeld, S.M., Quandt, R.E. & Trotter, H.F. (1966) Maximization by quadratic hill-climbing. *Econometrica*, **34**, 541–551.

[55] Goldfeld, S.M., Quandt, R.E. & Trotter, H.F. (1968) Maximization by improved quadratic hill-climbing and other methods. Econometrics Research Memo. No. 95, Princeton U.

[56] Goldstein, A.A. & Price, J.F. (1971) On descent from local minima. *Math. Computing*, **25**, 569–574.

[57] Graves, R.L. & Wolfe, P. (1963) eds., *Recent Advances in Mathematical Programming*. McGraw–Hill: New York.

[58] Greenstadt, J. (1966) A richocheting gradient method for nonlinear optimization. *SIAM J. Appl. Math.*, **14**, 429–445.

[59] Haarhoff, P.C. & Buys, J.D. (1970) A new method for the optimization of a nonlinear function subject to nonlinear constraints. *Computer J.*, **13**, 178–184.

[60] Hestenes, M.R. (1969) Multiplier and gradient methods. *J. Opt. Theory Applns.*, **4**, 303–320.

[61] Hestenes, M.R. & Stiefel, E. (1952) Methods of conjugate gradients for solving linear systems. *J. Res. Natl. Bureau Standards*, **49**, 409–436.

[62] Himmelblau, D.M. (1972) A uniform evaluation of unconstrained optimization techniques, in [72], 69–97.

[63] Huang, H.Y. (1970) A unified approach to quadratically convergent algorithms for function minimization. *J. Opt. Theory Applns.*, **5**, 405–423.

[64] Karlin, S. (1959) *Mathematical Methods and Theory in Games, Programming and Economics*, Vol. I. Addison–Wesley: Reading, Mass.

[65] Klingman, W.R. & Himmelblau, D.M. (1964) Nonlinear programming with the aid of a multiple gradient summation technique. *J. ACM.*, **11**, 400–415.

[66] Kuhn, H.W. (1970) ed., *Proceedings of the Princeton Symposium on Mathematical Programming*. Princeton U. Press: Princeton.

[67] Kuhn, H.W. & Tucker, A.W. (1951) Nonlinear programming, in [91], 481–492.

[68] Kuo, F.F. & Magnuson, W.M. (1969) eds., *Computer Oriented Circuit Design*. Prentice–Hall: Englewood Cliffs, N.J.

[69] Levenberg, K. (1944) A method for the solution of certain nonlinear problems in least squares. *Quart. J. Appl. Math.*, **2**, 164–168.

[70] Lill, S.A. (1970) A modified Davidon method for finding the minimum of a function using difference approximations for derivatives. *Computer J.*, **13**, 111–113.

[71] Lill, S.A. (1972) Generalization of an exact method for solving equality constrained problems to deal with inequality constraints, in [72], 383–393.

[72] Lootsma, F.A. (1972) ed., *Numerical Methods for Non-linear Optimization*. Academic: New York.

[73] Lootsma, F.A. (1972a) A survey of methods for solving constrained minimization problems via unconstrained minimization, in [72], 313–347.

[74] Luenberger, D.G. (1969) *Optimization by Vector Space Methods*. Wiley: New York.

[75] Luenberger, D.G. (1973) *Introduction to Linear and Nonlinear Programming*. Addison–Wesley: Reading, Mass.

[76] Maehly, H.J. (1963) Methods for fitting rational approximations. *J. ACM.*, **10**, 257–277.

[77] Mangasarian, O.L. (1973) Unconstrained Lagrangians in nonlinear programming. Computer Sci. Technical Report No. 174, U. of Wisconsin-Madison.

[78] Marquardt, D.W. (1963) An algorithm for least-squares estimation of non-linear parameters. *J. SIAM.*, **11**, 431–441.

[79] Matthews, A. & Davies, D. (1971) A comparison of modified Newton methods for unconstrained optimization. *Computer J.*, **14**, 293–294.

[80] Meinardus, G. (1967) *Approximation of Functions: Theory and Numerical Methods.* (translated by L.L. Schumaker) Springer: Berlin.

[81] McCormick, G.P. (1969) The rate of convergence of the reset Davison variable metric method. Math. Res. Center Report No. 1012, U. of Wisconsin-Madison.

[82] McCormick, G.P. (1969a) Anti zig-zagging by bending. *Managmt. Sci.*, **15**, 379–392.

[83] McCormick, G.P. (1970) The variable reduction method for nonlinear programming. *Managmt. Sci.*, **17**, 146–160.

[84] McCormick, G.P. (1971) Penalty function versus non-penalty function methods for constrained nonlinear programming problems. *Math. Prog.*, **1**, 217–238.

[85] McCormick, G.P. (1972) Attempts to calculate global solutions of problems that may have local minima, in [72], 209–221.

[86] McCormick, G.P. & Pearson, J.D. (1969) Variable metric methods and unconstrained optimization, in [34], 307–326.

[87] McCormick, G.P. & Ritter, K. (1972) Methods of conjugate directions versus quasi–Newton methods. *Math. Prog.*, **3**, 101–116.

[88] Murtagh, B.A. & Sargent, R.W.H. (1969) A constrained minimization method with quadratic convergence, in [34], 215–246.

[89] Murtagh, B.A. & Sargent, R.W.H. (1970) Computational experience with quadratically convergent minimization methods. *Computer J.*, **13**, 185–194.

[90] Nelder, J.A. & Mead, R. (1965) A simplex method for function minimization. *Computer J.*, **7**, 308–313.

[91] Neyman, J. (1951) ed., *Proceedings of the Second Berkeley Symposium on Mathematical Statistics and Probability*. U. of California Press: Berkeley, Cal.

[92] Oren, S.S. & Luenberger, D.G. (1974) Self-scaling variable metric (SSVM) algorithms I. *Managmt. Sci.*, **20**, 845–862.

[93] Oren, S.S. (1974) Self-scaling variable metric (SSVM) algorithms II. *Managmt. Sci.*, **20**, 863–874.

[94] Osborne, M.R. & Ryan, D.M. (1970) On penalty function methods for nonlinear programming problems. *J. Math. Anal. Applns.*, **31**, 559–578.

[95] Osborne, M.R. & Ryan, D.M. (1972) A hydrid algorithm for non-linear programming, in [72], 395–410.

[96] Palmer, J.R. (1969) An improved procedure for orthogonalising the search vectors in Rosenbrock's and Swann's direct search optimization methods. *Computer J.*, **12**, 69–71.

[97] Parkinson, J.M. & Hutchinson, D. (1972) A consideration of non-gradient algorithms for the unconstrained optimization of functions of high dimensionality, in [72], 99–114.

[98] Powell, M.J.D. (1962) An iterative method for finding stationary values of a function of several variables. *Computer J.*, **5**, 147–151.

[99] Powell, M.J.D. (1964) An efficient method for finding the minimum of a function of several variables without calculating derivatives. *Computer J.*, **7**, 155–162.

[100] Powell, M.J.D. (1965) A method for minimizing a sum of squares of non-linear functions without calculating derivatives. *Computer J.*, **7**, 303–307.

[101] Powell, M.J.D. (1966) Minimization of functions of several variables, in [127], 143–157.

[102] Powell, M.J.D. (1969) A method for nonlinear constraints in minimization problems, in [34], 283–298.

[103] Powell, M.J.D. (1970) A new algorithm for unconstrained optimization, in [119], 31–65.

[104] Powell, M.J.D. (1971) On the convergence of the variable metric algorithm. *J. Inst. Math. Applns.*, **7**, 21–36.

[105] Powell, M.J.D. (1971a) Recent advances in unconstrained optimization. *Math. Prog.*, **1**, 26–57.

[106] Powell, M.J.D. (1972) Some properties of the variable metric algorithm, in [72], 1–17.

[107] Powell, M.J.D. (1972a) Quadratic termination properties of a class of double-rank minimization algorithms. A.E.R.E. (U.K.) Harwell Technical Report No. HL72/456.

[108] Powell, M.J.D. (1972b) Some theorems on quadratic termination properties of minimization algorithms. A.E.R.E. (U.K.) Harwell Technical Report No. HL72/457.

[109] Powell, M.J.D. (1972c) Unconstrained minimization and extensions for constraints, A.E.R.E. (U.K.) Harwell Technical Report No. Hl72/3772.

[110] Remez, E. (1934) Sur le calcul effectif des polynomes d'approximation de Tchebychef. *Comt. Rend. Acad. Sci. (Paris)*, **199**, 337–340.

[111] Rhead, D.G. (1971) Some numerical experiments on Zangwill's method for unconstrained minimization. Inst. of Computer Sci. Working Paper No. 319, U. of London.

[112] Rhead, D.G. (1972) Further experiments on Zangwill's method. Inst. of Computer Sci. Working Paper No. 347, U. of London.

[113] Ritter, K. (1972) A superlinearly convergent method for minimization problems with linear inequality constraints. *Math. Prog.*, **4**, 44–71.

[114] Robinson, S.M. (1972) A quadratically-convergent algorithm for general nonlinear programming problems. *Math. Prog.*, **3**, 145–156.

[115] Rockafellar, R.T. (1973) Augmented Lagrange multiplier functions and duality in nonconvex programming. Mathematics Department Report, U. of Washington, Seattle.

[116] Rosen, J.B. (1960) The gradient projection method for nonlinear programming, I. Linear constraints. *J. SIAM.*, **8**, 181–217.

[117] Rosen, J.B. (1961) The gradient projection method for nonlinear programming, II. Nonlinear constraints. *J. SIAM.*, **9**, 514–532.

[118] Rosen, J.B. & Kreuser, J. (1972) A gradient projection algorithm for non-linear constraints, in [72], 297–300.

[119] Rosen, J.B., Mangasarian, O.L. & Ritter, K. (1970) eds., *Nonlinear Programming*. Academic: New York.

[120] Rosenbrock, H.H. (1960) An automatic method for finding the greatest or least value of a function. *Computer J.*, **3**, 175–184.

[121] Sargent, R.W.H. & Sebastian, D.J. (1972) Numerical experience with algorithms for unconstrained minimization, in [72], 45–68.

[122] Sayama, H., Kameyama, Y., Nakayama, H. & Sawaragi, Y. (1973) Generalized Lagrangian functions for mathematical programming problems. Dept. of Engineering Report, U. of Okayama, Japan.

[123] Schrager, R.J. (1970) Non-linear regression with linear constraints: An extension of the magnified diagonal method. *J. ACM.*, **17**, 446–452.

[124] Shah, B.V., Buehler, R.J. & Kempthorne, O. (1964) Some algorithms for minimizing a function of several variables. *J. SIAM.*, **12**, 74–92.

[125] Shanno, D.F. (1970) Conditioning of Quasi–Newton methods for function minimization. *Math. Computing*, **24**, 647–657.

[126] Stewart, G.W. (1967) A modification of Davidon's minimization method to accept difference approximations of derivatives. *J. ACM.*, **14**, 72–83.

[127] Swann, W.H. (1964) Report on the development of a new direct search method of optimization. ICI Central Instrument Lab. Research Note No. 64/3.

[128] Temes, G.C. (1969) Optimization methods in circuit design, in [68], 191–249.

[129] Walsh, J. (1966) ed., *Numerical Analysis: An Introduction.* Academic: London.

[130] Wilde, D.J. (1964) *Optimum Seeking Methods.* Prentice–Hall: Englewood Cliffs, N.J.

[131] Wilde, D.J. & Beightler, C.S. (1967) *Foundations of Optimization.* Prentice–Hall: Englewood Cliffs, N.J.

[132] Wilson, R.B. (1963) A simplicial method for convex programming. Ph.D. Dissertation, Harvard U.

[133] Wolfe, P. (1963) Methods of nonlinear programming, in [57], 67–86.

[134] Wolfe, P. (1967) Methods of nonlinear programming, in [1], 97–131.

[135] Wolfe, P. (1969) Convergence conditions for ascent methods. *SIAM. Rev.* **11**, 226–235.

[136] Wolfe, P. (1969a) Discussion of a review of a paper of D. Davies and R.H. Swann, in [34], 201–202.

[137] Zangwill, W.I. (1967) Nonlinear programming via penalty functions *Managmt. Sci.*, **13**, 344–358.

[138] Zangwill, W.I. (1968) Minimizing a function without calculating derivatives. *Computer J.*, **10**, 293–296.

[139] Zangwill, W.I. (1969) *Nonlinear Programming: A Unified Approach.* Prentice–Hall, Englewood Cliffs, N.J.

[140] Zoutendijk, G. (1970) Nonlinear programming, Computational methods, in [2], 37–86.

Further reading

Introductory texts

Box, M.J., Davies, D. & Swann, W.H. (1969) *Non-linear Optimization Techniques*, ICI Monograph No. 5, Oliver & Boyd: Edinburgh.

Dixon, L.C.W. (1972) *Nonlinear Optimization*. English Universities Press: London.

Wilde, D.J. (1964) *Optimum Seeking Methods*. Prentice–Hall: Englewood Cliffs, N.J.

Wilde, D.J. & Beightler, C.S. (1967) *Foundations of Optimization*. Prentice–Hall: Englewood Cliffs, N.J.

Kowalik, J. & Osborne, M.R. (1968) *Methods for Unconstrained Optimization Problems*. Elsevier: New York.

More advanced texts

Fiacco, A.V. & McCormick, G.P. (1968) *Nonlinear Programming: Sequential Unconstrained Minimization Techniques*. Wiley: New York.

Luenberger, D.G. (1972) *Introduction to Linear and Nonlinear Programming*. Addison–Wesley: Reading, Mass.

Zangwill, W.I. (1969) *Nonlinear Programming: A Unified Approach*. Prentice–Hall: Englewood Cliffs, N.J.

Conference proceedings

Graves, R.L. & Wolfe, P. (1963) eds., *Recent Advances in Mathematical Programming*. McGraw–Hill: New York.

Abadie, J. (1967) ed., *Nonlinear Programming*. North Holland: Amsterdam.
Kuhn, H.W. (1970) ed., *Proceedings of the Princeton Symposium on Mathematical Programming*. Princeton U. Press: Princeton.
Fletcher, R. (1969) ed., *Optimization*. Academic: New York.
Abadie, J. (1970) ed., *Integer and Nonlinear Programming*. North Holland: Amsterdam.
Rosen, J.B., Mangasarian, O.L. & Ritter, K. (1970) eds., *Nonlinear Programming*. Academic: New York.
Lootsma, T.A., (1972) ed., *Numerical Methods for Non-linear Optimization*. Academic: New York.
Andersson, R.S., Jennings, L.S. & Ryan, D.M. (1972) *Optimization* U. of Queensland Press: St. Lucia, Queensland.

Special topics

Theory

Dempster, M.A.H. (1975) *Elements of Optimization*. Chapman & Hall: London.
Luenberger, D.G. (1969) *Optimization by Vector Space Methods*. Wiley: New York.

Linear programming

Gale, D. (1960) *The Theory of Linear Economic Models*. McGraw–Hill: New York.
Gass, S.I. (1969) *Linear Programming*. 3rd Edition, McGraw–Hill: New York.
Kunzi, H.P., Tzshach, H.G. & Zehndele, C.A. (1968) *Numerical Methods of Mathematical Optimization with ALGOL and FORTRAN Programs*, Academic: New York.

Large systems

Lasdon, L.S. (1970) *Optimization Theory for Large Systems*. Macmillan: London.

Approximation

Fox, L. & Parker, I.B. (1968) *Chebyshev Polynomials in Numerical Analysis*, Oxford U. Press: London.
Meinardus, G. (1967) Approximation of Functions: Theory and Numerical Methods. (translated by L.L. Schumacher) Springer: Berlin.

Solution of nonlinear equations

Ortega, J.M. & Rheinboldt, W.C. (1970) *Iterative Solution of Non-linear Equations in Several Variables*. Academic: New York.

Applications and applications oriented texts

Cottle, R.W. & Krarup, J. (1973) eds., *Optimization Methods for Resource Allocation*. English Universities Press: London.

Fox, R.L. (1971) *Optimization Methods for Engineering Design*. Addison–Wesley: Reading, Mass.

Goldfeld, S.M. & Quandt, R.E. (1972) *Nonlinear Methods in Econometrics*, North Holland: Amsterdam.

Institution of Electrical Engineers (1970) *Optimization Techniques in Circuit and Control Applications*. Conference Publication No. 66.

Leitmann, G. (1962) ed., *Optimization Techniques with Application to Aerospace Systems*, Academic: New York.

Mickle, M.H. & Aze, T.W. (1972) *Optimization in Systems Engineering*. Intext: Scranton, Penna.

Nicholson, T.A.J. (1971) *Optimization in Industry*. 2 Vols. Longman: London.

Index